The Microsoft™ Excel Supplement

MARK DUMMELDINGER

STATISTICS
for Business and Economics

EIGHTH EDITION

McCLAVE ▪ BENSON ▪ SINCICH

Upper Saddle River, NJ 07458

Acquisitions Editor: Kathy Boothby Sestak
Supplement Editor: Joanne Wendelken
Special Projects Manager: Barbara A. Murray
Production Editor: Wendy A. Perez
Supplement Cover Manager: Paul Gourhan
Supplement Cover Designer: PM Workshop Inc.
Manufacturing Buyer: Lisa McDowell

© 2001 by Prentice Hall
Upper Saddle River, NJ 07458

All rights reserved. No part of this book may be reproduced, in any form or by any means, without permission in writing from the publisher.

Printed in the United States of America

10 9 8 7 6 5 4 3 2 1

ISBN 0-13-029347-4

Prentice-Hall International (UK) Limited, London
Prentice-Hall of Australia Pty. Limited, Sydney
Prentice-Hall Canada, Inc., Toronto
Prentice-Hall Hispanoamericana, S.A., Mexico
Prentice-Hall of India Private Limited, New Delhi
Pearson Education Asia Pte. Ltd., Singapore
Prentice-Hall of Japan, Inc., Tokyo
Editora Prentice-Hall do Brazil, Ltda., Rio de Janeiro

Table of Contents

Primer   **Excel Basics Needed for Statistical Analysis of Data**

- **P.1**   **Introduction and Overview**   1
  - P.1.1   Versions of Excel   1
  - P.1.2   Versions of Windows   1
  - P.1.3   What May be Skipped   1
  - P.1.4   More Detailed Information on Excel   2

- **P.2**   **What You Need to Know to Begin Using Excel**   2
  - P.2.1   Using the Mouse   2
    - P.2.2.1   Starting and Exiting PHStat   3
    - P.2.2.2   Starting and Exiting Excel   3
  - P.2.3   Layout of Worksheets and Worksheets   5
  - P.2.4   Menus, Toolbars and Dialog Boxes   7
  - P.2.5   Manipulating Windows   9

- **P.3**   **Ways to Get Help**   10
  - P.3.1   Help on the Main Menu   10
  - P.3.2   Answer Wizard   12

- **P.4**   **Opening and Saving Documents**   12
  - P.4.1   Opening a Brand New Spreadsheet   12
  - P.4.2   Opening a File you have Already Created   12

- **P.5**   **Entering Information**   13
  - P.5.1   Activating a Cell or Range of Cells   13
  - P.5.2   Types of Information   14
  - P.5.3   Changing Information   14
  - P.5.4   Moving and Copying Information   15

- **P.6**   **Formatting Numbers**   16
  - P.6.1   Aligning Information   17
  - P.6.2   Formatting a Range   17
  - P.6.3   Inserting or Deleting Rows and Columns   17
  - P.6.4   Filling Adjacent Cells   18
  - P.6.5   Series   19
  - P.6.6   Sorting   19

- **P.7**   **Saving and Retrieving Information**   21
  - P.7.1   Naming Workbooks   21

- **P.8**   **Printing**   22
  - P.8.1   Page Setup   22
  - P.8.2   Page   22
  - P.8.3   Margins   23
  - P.8.4   Header/Footer   23

	P.8.5	Sheet 24
	P.8.6	Enhancing Output 25
	P.8.7	Inserting and Removing Page Breaks 25
	P.8.8	Preview and Print 25

P.9   **Using Formulas and Functions**   **26**
- P.9.1 Operators 26
- P.9.2 Order of Operators 27
- P.9.3 Writing Equations 27

P.10   **Entering Formulas**   **29**
- P.10.1 Relative References 30
- P.10.2 Absolute References 31

Chapter 1   **Statistics, Data, and Statistical Thinking**   **33**
- 1.1 Introduction 33

Chapter 2   **Methods for Describing Sets of Data**   **35**
- 2.1 Introduction 35
- 2.2 Graphical Techniques in Excel 36
  - 2.2.1 Bar Graphs and Histograms 36
  - 2.2.2 Pie Charts 42
  - 2.2.3 Scatter Plots 44
  - 2.2.4 Stem-and-Leaf Displays 49
  - 2.2.5 Box Plots 50
- 2.3 Numerical Techniques in Excel 51
  - 2.3.1 Measures of Center 51
  - 2.3.2 Measures of Spread 54
  - 2.3.3 Measures of Relative Standing 55
- Technology Lab 58

Chapter 3   **Probability**   **63**
- 3.1 Introduction 63
- 3.2 Probabilities in a 2x2 Table 63
- 3.3 Random Sampling 65

Chapter 4   **Discrete Random Variables**   **69**
- 4.1 Introduction 69
- 4.2 Calculating Binomial Probabilities 69
- 4.3 Calculating Poisson Probabilities 72
- Technology Lab 74

Chapter 5   **Continuous Random Variables**   **75**
- 5.1 Introduction 75
- 5.2 Calculating Normal Probabilities 75
- 5.3 Assessing the Normality of a Data Set 78
- 5.4 Calculating Exponential Probabilities 80
- Technology Lab 82

Contents v

Chapter 6   **Sampling Distributions**   83
        6.1   Introduction   83
        6.2   Calculating Probabilities Using the Sampling Distribution of the Sample Mean   83
    Technology Lab   85

Chapter 7   **Inferences Based on Single Sample: Estimation with Confidence Intervals**   87
    7.1   Introduction   87
    7.2   Estimation of a Population Mean - Sigma Unknown   87
    7.3   Estimation of a Population Proportion   91
    7.4   Determining the Sample Size   93
        7.4.1   Determining the Sample Size for Means   93
        7.4.2   Determining the Sample Size for Proportions   94
    Technology Lab   96

Chapter 8   **Inferences Based on a Single Sample: Tests of Hypothesis**   99
    8.1   Introduction   99
    8.2   Tests of Hypothesis of a Population Mean - Sigma Unknown   99
    8.3   Tests of Hypothesis of a Population Proportion   103
    Technology Lab   106

Chapter 9   **Inferences Based on Two Samples: Confidence Intervals and Tests of Hypothesis**   109
    9.1   Introduction   109
    9.2   Tests for Differences in Two Means   109
    9.3   Tests for Differences in Two Proportions   112
    9.4   Tests for Differences in Two Variances   115
    Technology Lab   117

Chapter 10   **Simple Linear Regression**   121
    10.1   Introduction   121
    10.2   The Coefficient of Correlation   121
    10.3   The Coefficient of Determination and Regression Output   123
    10.4   Estimating and Predicting with a Simple Linear Model   126
    Technology Lab   128

Chapter 11   **Multiple Regression and Model Building**   131
    11.1   Introduction   131
    11.2   Multiple Regression Model Building   131
    11.3   Comparing Two Regression Models   140
    11.4   Stepwise Regression   141
    11.5   Residual Analysis   142
    Technology Lab   150

Chapter 12   **Methods of Quality Improvement**   153
    12.1   Introduction   153
    12.2   Constructing R-Charts and $\bar{x}$-Charts   153
    12.3   Constructing p-charts   156
    Technology Lab   160

Contents

Chapter 13   **Time Series: Descriptive Analyses, Models, and Forecasting**   163
       13.1   Introduction   163
       13.2   Descriptive Analyses: Index Numbers   164
       13.3   Exponential Smoothing   170
       13.4   Using Regression to Model Time Series Data   172

Chapter 14   **Design of Experiments and Analysis of Variance**   175
       14.1   Introduction   175
       14.2   The Completely Randomized Design   175
       14.3   The Factorial Design   178
       14.4   Using Regression Analysis for ANOVA   180

Chapter 15   **Nonparametric Statistics**   181
       15.1   Introduction   181
       15.2   The Wilcoxon-Rank Sum Test for Independent Samples   181
       15.3   The Kruskal-Wallis H-Test for the Completely Randomized Design   182
       Technology Lab   184

Chapter 16   **Categorical Data Analysis**   187
       16.1   Introduction   187
       16.2   Testing Categorical Probabilities: Two-Way Table   187
       Technology Lab   191

Primer
Excel Basics Needed for Statistical Analysis of Data

P.1 Introduction and Overview

This manual is designed for use with McClave, Benson, and Sincich *Statistics for Business and Economics*, 8th Edition. It is not intended to take the place of your Excel user's manual, however it will introduce the Excel novice to the software and provide the basic tools necessary to analyze statistical data using Excel. To accomplish this goal, we sue a four-part process. First, we introduce the statistical procedures available in Excel. Next, we illustrate these procedures by teaching you how to perform the Excel commands required to produce the output from selected examples in Statistics for Business and Economics 8/e. Both the steps taken and output generated are provided in this manual to teach you the Excel steps to be followed. Third, we have provided an Excel Lab where the student is given Excel workbooks to use in answering step-by-step questions. The final part of our process is to provide you with Excel data sets that may be used to complete homework exercises in the McClave/Benson/Sincich text. Our hope is that this "introduce-learn-practice" format will enable you to finish the course with a firm understanding of how Excel can be used to analyze statistical data.

We believe that the Excel portion of a statistics course should serve strictly to enhance the statistics that is being taught. We have tried to keep this philosophy in mind when writing this manual. We have attempted to provide an easy-to-use format that will allow you to use Excel to calculate the statistics you learn in class. If we have been successful, you will view Excel as a valuable tool for the statistician. Used correctly, Excel allows the statistician to spend more time using, and less time calculating, the kind of information that you will explore in your statistics course.

P1.1 Versions of Excel

This manual principally uses Excel 97 (version 8.0). Many different versions of Excel exist; Excel 97 is the most prevalent at the time of this writing. All of the versions function in essentially the same way, whether used on the PC or a Mac, but you will notice slight differences in how the screens look and in the names of some commands. In this manual, we will primarily use the statistics procedures available in the PHStat add-in currently available from Prentice Hall for Excel. When possible, we have supplemented these procedures with different Functions and techniques available within the Excel program. Our goal throughout this manual is to provide the user with the easiest method of generating the desired output.

P.1.2 Versions of Windows

The copies of screens shown in this book are taken from a PC using Windows 98. They will appear slightly different if using Windows 3.x or a Mac operating under System 7. After you are operating comfortably within Excel, these differences should be minor. There will, however, be slight differences between Macs and PCs in the keys used for commands.

P.1.3 What may be skipped

If you have used spreadsheets before, you can probably omit much of the first chapter. Other programs, such as Lotus 1-2-3 and Quattro, use slightly different terminology when describing the spreadsheets. The concepts, however, are essentially the same although the terms and/or procedures may differ slightly.

- For those of you who are looking for commands for a specific procedure or example from *Statistics for Business and Economics*, 8/e you will find that each of the chapters may be used independently.

P.1.4 More detailed information on Excel

A glance at the bookshelves in the computer section of most bookstores will reveal a number of books that deal with Excel in all its various versions. Few deal with Excel as a way to perform statistical analyses.

Books we have referred to when preparing this manual include the following:

Dretzke, B., & Heilman, K. (1998). *Statistics with Microsoft® Excel*. Upper Saddle River, NJ. Prentice Hall

Gold, L. & Post, D. (1995). *The Complete Idiot's Guide to Excel for Windows 95*. Indianapolis, IN: Que.

Halvorson, M. & Young, M. (1997). *Running Microsoft Office 97*. Redmond, WA: Microsoft Press.

Marmel, E., Bucki, L., & Guilford, E. (1995). *The Big Basics book of Excel for Windows 95*. Indianapolis, IN: Que.

Microsoft Corporation (1994). *User's Guide. Microsoft Excel*. Redmond, Wa: Microsoft Press.

Nelson, S.L. (1996). *Microsoft Excel 97 Field Guide*. Redmond, WA: Microsoft Press.

Neufeld, J.L. (1997). *Learning Business Statistics with Microsoft Excel*. Upper Saddle River, NJ: Prentice Hall.

Nicholson, J.R. & Nicholson, S.R. (1997). *Discover Excel 97*. Foster City, CA: IDG Books.

P.2 What You Need to Know to Begin Using Excel

P.2.1 Using the mouse

Mice come in several forms. The majority are provided with new computers and roll on the desktop or pad. A small ball on the bottom, when rolled, causes the pointer on the screen (called the **screen pointer**) to move in a corresponding way. Another version (called a trackball) places a larger ball in a framework that allows you to roll the ball with your fingers. Finally, there are other forms that have small screens that you move your fingers across as you would move the mouse. The pressure of your finger moving across the screen causes a screen pointer to move in synchrony with your movements.

All devices have at least one, and most likely two or more, buttons that you can click or hold down, sometimes while also moving the ball. There are four basic actions you will need to use in operating the mouse:

- **Point** You point to objects on the screen by sliding the mouse on the deskpad or rolling the trackball. The screen pointer will track the movements made on your desk. The shape of the screen pointer will change, most often being an outline arrow or the outline of a plus sign when using Excel, but changing with the task to be done.

- **Click** "Click" means to press and release the left mouse button (called a left-click). If you are pointing at an executable command, this action will cause it to take place. If you point to any cell on the spreadsheet and click, that cell becomes the **active cell** and is ready to receive data. Sometimes you may be asked to press and release the right mouse button (called a right-click), which is commonly used to place a shortcut menu on the screen. The above assumes you are using the settings for the mouse provided by the manufacturer with your right hand. If you are left-handed or want to reverse the way the buttons function, this can be done. Click on the **Help** icon above the **Start** button on the Windows 95 screen and look for *mouse, buttons, reversing* in the **Index**.

- **Double-click** "Double-click" means to press and release the left mouse button twice rapidly. If you fail to press rapidly enough, it is interpreted as one click. Often this process replaces the two-step sequence of selecting a command and then clicking on **OK** to execute that command.

- **Drag.** Objects on the screen are moved by dragging. To drag, place the mouse pointer on the item you want to move, click and hold the left mouse button --- do NOT release it. While you hold the mouse button down, slide the mouse to move the screen pointer and the item to the location you want. Then release the mouse button.

P.2.2.1 Starting and Exiting PHStat

To start the program using either Windows 95 or 98:
Click on **Start** in the lower left of the screen. Move the mouse to **Programs** and then continue moving through menus until you find the **PHStat** icon to click and begin. Click on **Enable Macros** and **Continue** to open the **PHStat** program. You are now ready to begin.

To end the program using either Windows 95 or 98:
Click in the upper **X** (the **close button**) that you find in the upper right corner of the screen. (If two sets of boxes are showing, the lower set applies to the **worksheet** (spreadsheet) you have showing while the upper one is for the application or program itself (i.e., PHStat). IF you have edited (changed) any of the information in the workbook, you will be prompted to save the information before closing the program. In many computer labs, you will be asked to save all of your data on a diskette. We will describe this procedure later.

P.2.2.2 Starting and exiting Excel

To start the program using either Windows 98 or 95:

Windows 98 or 95: Click on **Start** in the lower left of the screen. Move the mouse to **Programs** and then continue moving through menus until you find the **Microsoft Excel** icon to click and begin.

4 Primer: Excel Basics

If you have Office 95 or Office 97, you may have to find the folder containing the suite, open it, and double-click on the icon of the Excel program you find there.

You may want to go through the Quick Preview online tutorial if you are unfamiliar with spreadsheets and need a quick overview. This is often found in the Excel folder.

Windows 3.x: In the **Program Manager**, locate the icon or folder containing the Excel icon. Double-click on it to begin.

To <u>exit</u> the program using either Windows 95 or 3.11:

Windows 98 or 95: Click in the upper **X** (the **close button**) that you find in the upper right corner of the screen. (If two sets of boxes are showing, the lower set applies to the worksheet (spreadsheet) you have showing while the upper one is for the application or program itself, (i.e., Excel)). If you have edited (changed) any of the information in the **workbook**, you will be prompted to save the information before closing the program. In many computer labs, you may be asked to save all of your data on a diskette. We will describe this procedure later.

Windows 3.x: Use the same procedure as above. The **X** is a general sign to indicate you want to stop using or to exit a program. You can also select **Exit** from the **File** menu for either Windows 95 or 98 or Windows 3.x.

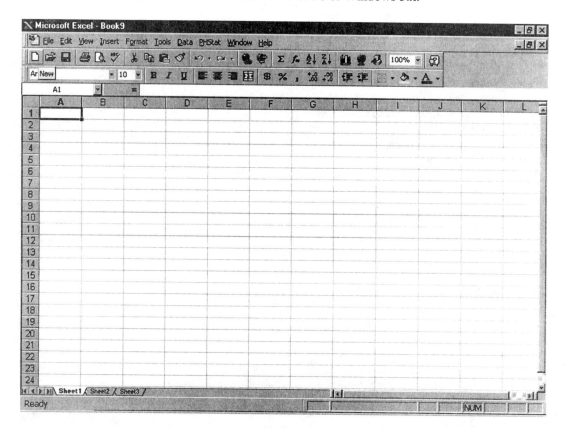

P.2.3 Layout of worksheets and workbooks

The above figure shows what is on the screen when PHStat for Excel 97 is opened. It may be slightly different from what you have on your screen. The list that follows briefly describes each of the items on the figure above, starting at the top.

- **Program Title Bar** This is most likely at the very top of the screen, with the default title being Microsoft Excel - Book 1. It indicates the name of the application and the fact that you are in what is called *Book 1*, the name given to a newly opened spreadsheet. Each book initially consists of 16 worksheets (this can be changed) which are stored together as a unit called a book. When you save your work, all of the sheets in this book will be saved together as one file.

- **Program Icons** Known as **sizing buttons** in Office 97, these are at the very right on the Program Title Bar, just as on the screen shown above. There are three program icons. Each is described below.

    - The **minimize button** shrinks the program which is then represented as a button on the taskbar at the bottom of the screen. This is part of Windows 95, which inactivates, but does not close, programs and places their icon at the bottom of the screen so they can be immediately reactivated.

    - The **maximize button** switches (toggles) between the full-screen and window views.

    - The **close button,** which closes the Excel application. You will be asked if you want to save your work, if there is any, before the application is closed.

- **Main menu bar.** Commands are grouped into categories such as *File*, *Edit*, *View*, etc. Clicking on one of these will drop down the commands in that group. Those that are useful for our purposes will be discussed later.

- **Workbook Icons.** These are the same as the **program icons** described above, except that they apply to the specific workbook being used rather than to the program.

- **Standard Toolbar.** This is a ribbon of icons that are designed to ease your access to commands. Some are not as obvious as one might wish, but all you need to do to find out what the button does is to point at it with your mouse pointer and pause. Its name is then displayed. For example, the first icon, labeled **New**, resembles a sheet of paper and will create a new workbook if it is clicked. The second will open a file, etc. You can already see how there are a number of ways to execute the same command. Near the extreme right of the toolbar we have the **Tip Wizard** (like a light bulb) which will give us information about how to do something more effectively. Click on it and you will see a general tip inserted in a space between the **Formatting Toolbar** and the **Formula Bar**. Click again to remove it. Next to Tip Wizard is another form of *Help*, the **Help Tool**. If you click on that button, the screen pointer changes to a question mark. Move it to the part of the screen you need to know more about, click the mouse, and it will give you a list of topics to explain the specific question you have. Try it by clicking on it and then moving to the letters at the top of each column. You will be given a list of *Parts of the Microsoft Excel Screen* that you can choose from for a brief explanation of their purpose. Click on *Column Heading* and you will be given a brief explanation and more options, such as adjusting column width, hiding a row or column, etc.

- **Formatting Toolbar.** Commands that change the appearance of your text are found here. First, you have the default font listed, *Arial* in most cases. If you wish to use a different font you need only click on the down arrow next to the font box to see which ones are available. To change the default font for the whole spreadsheet, use **Help** and search for the term *fonts*. Click on the second option to obtain a *How To* box with directions for changing the default startup workbook or the standard font.

The size of the font (10) is displayed in the next window with an arrow for changing it.

Following this are buttons for **bold**, *italic*, and underline, left-alignment, center-alignment, and centering across columns. The next five icons control the form of currency, percentage, use of commas in numbers, and the number of decimals displayed. To place any of a variety of borders around a cell or cells, use the next icon. The last two deal with colors, which we will not cover.

- **Formula Bar.** The first window shows the address of the active cell in the worksheet that is displayed. Initially it is A1, so this is the address displayed. Move to a different cell and click on it to make it the active cell. Now that cell's address is displayed.

This first box on the formula bar is also called the **Name Box**. To learn how to use it, click on *Help* on the Main Menu Bar, then *Search for Help on* ..., and type in *name box* followed by clicking on *Display*. You will have an explanation of how to name a cell, formula, or range of cells.

Notice what happens to the formula Bar when we type some numbers in cell A1. Three more buttons appear:

- The red X is clicked when we want to "destroy" or delete the information we have typed in the active cell. That information is also displayed in the Formula Bar and will be removed from both sites is we click on the X.

- The green check mark, when clicked, indicates that the data as entered are acceptable. The data will remain in the cell, but the three buttons disappear indicating that the cell is not being edited.

- Finally, the f_X is an icon to turn on the **Function Wizard**, a set of over 300 functions in categories such as Financial, Math & Trig, Statistical, etc. We will use the statistical functions quite often, and discuss the use of this tool throughout the chapters.

- **Worksheet Area.** Finally, we have the worksheet, which consists of 4,194,304 cells with columns labeled as letters and rows as numbers. There are 256 columns and 16,384 rows. Press CTRL+DOWN ARROW (at the same time) to move to the last row and CTRL+RIGHT ARROW to move the last column. Each cell is identified by the combination of its column letter and row number as is displayed in the Name Box.

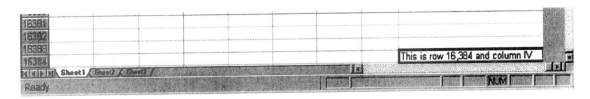

- **Scroll Bars** are found at the right and at the bottom of the worksheet. There are two small arrows, which look like a triangle laid on its side at either end of each scrollbar that, when clicked with the mouse, will move the screen up, down, right, or left one line for each press. Some users call the box within the scroll bar an *elevator*. You can grab the box or elevator and drag it. The screen will move a distance that corresponds to the amount you move the box. Finally, if you click in the shaded area between the box and the top or bottom of the scroll bar, the screen will move one whole screen in the direction you click. Thus, you can move a line at a time, a screen at a time, or from the top to the bottom of the screen. If you drag the elevator to the bottom, you will not move to 16,384 but only down about two screens. If you had data that extended to row 16,384, then the elevator would move to that location. Also, if you hold down the shift key while dragging the box, you can move to the end of the row or column.

- **Worksheet Tabs.** These are at the lower left of the screen and are labeled as *Sheet 1*, *Sheet 2*, etc. The default setting provides 16 of these tabs, which you can move through by clicking the arrows to the left of the name. Try it. Two arrows will move the active sheet to the left: one moves a sheet at a time and the other will move to the left-most sheet. The same applies to the right arrows. All 16 of these worksheets are stored together as one unit, called a Book.

P.2.4 Menus, toolbars, and dialog boxes

The Windows icon is at the very top of the screen to the left of Microsoft Excel. Use this to restore, minimize, close, etc.

The **Main Menu Bar** is at the top of the Excel screen. Click on *File* and you will have a set of commands drops down, beginning with *New* and ending with *Exit*. Most of these deal with operations on files. Notice that you can use combinations of keys to execute many of these commands. Key combinations are shown after the name. If you want to open a new file, press **CTRL+N** and you will have a new blank worksheet opened. **CTRL+V** will paste material from the **Clipboard** into the location you specify with the active cell.

Click somewhere on the blank spreadsheet and the dropdown commands will disappear. Now move the mouse pointer to the icons below the words that make up the Main Menu. These icons are on the Workbook Icon Bar. Hold the pointer to the second icon from the left for a moment and you will see the work *Open* appear. Click on that icon and you will open a new worksheet.

8 Primer: Excel Basics

Rather than reading our explanation of each command, here is how you can find out what each does: Move the mouse pointer to the up-arrow question mark at the extreme right side of the Workbook Icons. This is called the **Shortcut Help Icon** or **Help Tool**. Click on it Now the mouse pointer changes to a similar icon on the screen (the arrow is now in outline form). Move the pointer to **Edit** on the **Main Menu Bar**. Click so that you drop down the set of commands. Now click on **Copy**. Microsoft Excel Help now displays this box:

> **Copy (Edit menu)**
> Copies the selection to the Clipboard.

As you move to higher versions of Excel, the explanations become more elaborate. In Version 8, a longer description is given and some of the phrases are colored green and underlined. If you click on them you will be transferred to another screen that provides more detailed information.

Try clicking on the icon of a light bulb to the left of the Shortcut Help Icon. The Tip Wizard will give you hints on how to do things more efficiently.

The **Status Bar** is at the bottom of the Excel screen. (At the very bottom of the complete screen is the **Taskbar** for Windows 95 or 98). The Status Bar indicates what is happening. Most of the time, when nothing special is being done, it simply says *Ready* at the left of the screen. When you are editing a cell (changing information in it) then the word **Edit** will be displayed. Enter a number in cell A1, say 12359. Click the green check mark to accept the entry. The number is displayed in A1 and also in the Formula Bar. The three icons described above also disappear. If we want to change the 5 to a 6, move the pointer to the numbers in the formula bar. Now the pointer changes to an I-beam, which you can place just to the right of the 5, before the 9. Do this by clicking in that location. The three icons reappear because you have started to edit the contents of the cell. The word *Edit* is displayed in the Status Bar. Now press the backspace key and the 5 disappears. Enter a 6 and then press the Enter key or click on the check mark to accept the change.

Also shown at the right of the Status Bar is whether the **Number Lock** is turned on so you can use the keypad to enter numbers. Press the **Num Lock** key on the keyboard to turn it on or off. If the **Scroll Lock** is on or the **Caps Lock** keys are on this will also be shown here.

Dialog Boxes usually require that you choose from a number of alternatives. For example, in the **Save As** dialog box show below, you first must choose the place to save the file. Here we have the A: drive selected as the location. If you click on the arrow to the right, the set of possible locations will be displayed, including the C: drive and any servers that are available. You can then click on your choice to activate it and, if there are currently files of the same type in that location, they will be displayed. Hold the pointer to the icons to the right of the window for their name. A description of their function can be found in the Help file. **File name** is the place where you type the name you wish to use for the file. The default of *Book 2* is shown here. The type of file can be selected in the window below. In most cases, the default shown here is what you will use. Note the numerous options for the format in which a file can be saved when you click on the arrow by the window.

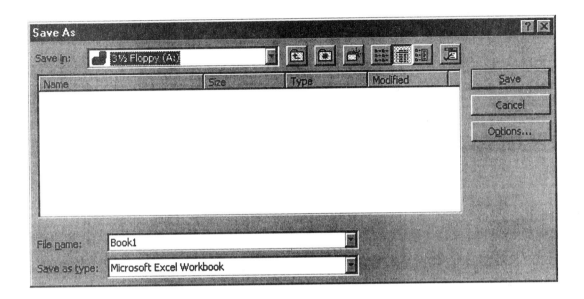

P.2.5 Manipulating Windows

We will discuss moving from one window to another, using the sizing buttons, and changing the size of windows using the mouse pointer.

The name **Windows** aptly describes the major feature of computers today. Excel is described as working within Windows 95 or 98. With windows, the user has the ability to have a number of programs or applications operating at the same time and displayed on the screen at the same time. You also have this opportunity within Excel, in that you can move from sheet to sheet within a workbook as well as changing to another workbook. All windows, as all sheets, are readily available, but only one is active and ready for use at a time. To move from one sheet to another in a workbook, simply click on the Sheet Tab at the bottom of the spreadsheet. As you move to Sheet 5, by clicking on it, Sheet 6 tab appears, and so on unitl you have moved through 16 sheets. You can also do this one sheet at a time with the center arrows to the left of the sheet tabs. Click on the outer two arrows to move to the extreme right (Sheet 16) or extreme left (sheet 1) of the series in that workbook. Instead of seeing Sheet 1, Sheet 2, etc., you can name these, say as Homework 1, Lab 2, etc. Use your knowledge of the Help option to find you how. As always, we recommend that you use the Shortcut Help Icon to clear up any questions about what different icons are called or how they are used. You may also need to add more sheets, which can be done by using the **Insert - Worksheet** command.

Sizing Buttons were described previously. In both the Excel program and in the Windows program these are displayed at the upper right. The Minimize Button reduces the spreadsheet to a small rectangle and places it near the bottom of the screen. When a spreadsheet has been minimized, a new icon appears, the Restore button. When click on, it will restore the spreadsheet to the size and location it had before being minimized. The Maximize Button enlarges the spreadsheet to fill all available space on the Screen. When you have minimized a spreadsheet (or a program in Windows) you have not closed or quit the program. It is still ready to go, but waiting in the wings, so to speak. Finally the **Close Button** resembles the X seen in the **Formula Bar** when editing data. This commands the program to close the book you are working on. You are then asked if you want to save changes you have made. If you say yes, then you are asked for the name to use in saving the workbook and location for it. Try it. When you have closed a file (Book in this case) you haven't

10 Primer: Excel Basics

stopped Excel from operating. The toolbars are still there but the Menu Bar now only has *File* and *Help* available. It is expected that you will create a new file or open an existing file.

P.3 Ways to Get Help

P.3.1 Help on the Main Menu

The Help command is the rightmost command on the Main Menu Bar. Click on it and you will see a menu providing these options: *Contents*, *Search for Help on ...*, and *Index*. These options may differ slightly with the version of Excel you are using. *Contents* presents broad categories which are broken down into finer sections. *Search* allows you to move through a list of topics and select the one most related to your problem. *Index* presents the topics as you would find them listed in the index at the back of a book. Within each of these you always have the option of moving from one to the other using commands near the top of the screen.

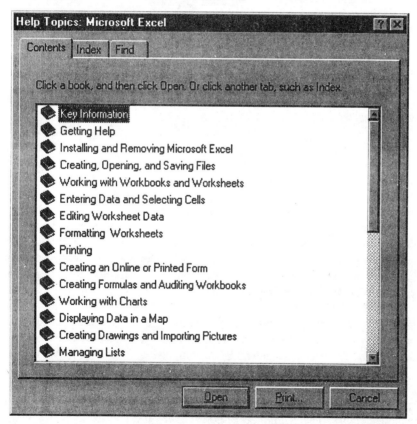

For the person with a specific question, the *Search* option is probably the most useful. You do, however, need to use the terms that the Help database uses. For example, assume that you wanted to change the format of numbers so that there are more or fewer decimals displayed. You begin by typing the word *numbers*, in the window. As you type, the list of terms moves to the word you type, followed by subcategories of that term. If you click *Display*, then more choices are given. Click on the one most related to your problem, and a description of how to deal with it will most likely appear. An example is given at the top of the next page.

P.3: Ways to Get Help 11

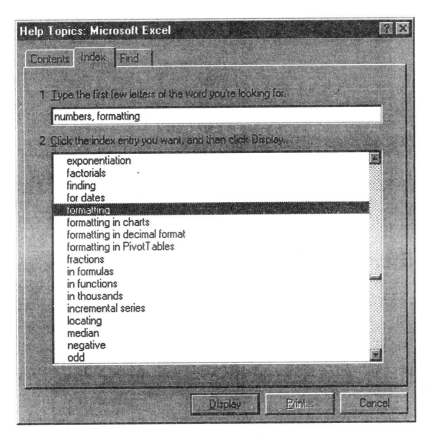

After you click the *Display* key on the above window, a smaller window with a finer breakdown of the topic appears:

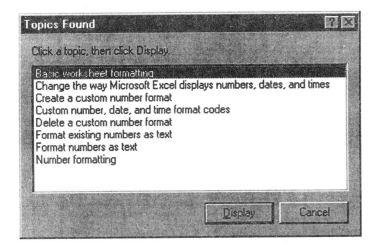

As you scroll down the list and select the appropriate description of the topic you want more information on, you may find that when you ask for help you will have the Answer Wizard window displayed. Instead of giving you written directions, it will give you a series of steps to follow on your spreadsheet that will answer your question.

P.3.2 Answer Wizard

The most recent versions of Excel provide this amusing, active image that will answer your questions. Artificial intelligence technology has been integrated into this feature, so that when you write a question you should get an appropriate response or set of selections that are likely to help you with your problem. It is a *context-sensitive* feature. What you are working on establishes the context, within which the Wizard will search. For example, if you are drawing a chart and need help in labeling the axes, the program recognizes that you are using this part of the program and gives a list of topics that are likely to relate the question you have. Click on one and you will be given more detailed help.

P.4 Opening and Saving Documents

1.4.1 Opening a brand new spreadsheet

If you started Excel by clicking the icon, the screen opened with a blank spreadsheet. The title at the top indicates it is called *Book 1* and the tabs across the bottom are labeled *Sheet 1, Sheet 2*, etc. Initially there are 16 sheets available. All sheets together are stored in one unit, called a book. You might, for example, choose to keep all homework assignments for one class in one unit, now called *Book 1*, but renamed by you as *Stats 1 HW*. You could then create other books that contain your personal budget, your records as the treasurer of an organization, and so on.

When you enter information into the worksheet, it is stored in the active memory of the computer, and will disappear is power is lost, whether by an accident that deprives you of electrical power or by turning off the machine without saving your work.

> It is wise to save your work often, especially if you are working on a complex project that requires many hours of data entry or processing. Use the automatic save feature. Find directions for using it in **saving documents, protecting work,** in the list of **Help** topics!

If you click on that X in the upper right of the Excel window, indicating that you want to quit, the program will ask you if you want to save the file you have created. If you have already saved it, and therefore have named it, Excel will save it using the same name unless you use the Save As ... command, which is used to change the name or location for saving.

P.4.2 Opening a file you have already created

If you want to continue working on a file that you previously created, you can just double-click on the icon that represents that file. The program will automatically open Excel when you open a file created by Excel. Sometimes this causes a problem if you create a file in Version 7 and want to open it using a computer that only contains Version 5. Files are often incompatible with earlier versions and cannot be opened by an earlier version. One solution is to use **Save As...** and save the file you created in version 7 in the format that can be used by version 5.

P.5 Entering Information

In the previous chapter we described the different parts of the screen – the Main Menu Bar, the Program Title Bar, program icons, the Standard Toolbar, the Formatting Toolbar, worksheet tables, Status Bar, and Scroll Bars. We also shoed the various shapes that the mouse pointer can take. Now we are ready to enter, save, edit, retrieve, and perform other data manipulations.

- **Addresses** Each cell is identified by a combination of a letter and a number to locate it in the spreadsheet. Letters for each column are shown across the top of the worksheet and begin to repeat with two letter combinations after Z is reached. This continues until the final column is reached, which is the 256th column, labeled IV. Rows continue numerically until the last row, 16,384, is reached.

P.5.1 Activating a Cell or Range of Cells

When the spreadsheet is initially opened, Cell A1 is automatically the active cell. It has a dark outline around it, which indicates that whatever you type will be entered into that cell. Note that the address A1 is displayed in the Name Box, which is to the left of the Formula Bar. Move to another cell, click, and note the change in the Name Box.

Often we need to refer to more than one cell at a time. A group of cells is called a **Range**. Click on cell B4, hold the mouse button down and drag down to B8. Release the button. The screen is now darkened (highlighted) in the range of cells, except for the top cell, B4. To indicate the address of a range of cells, we separate the addresses of the upper left and lower right cells with a colon. Here we have a range identified as B4:B8, although the Name Box only indicates the address of the top cell.

14 Primer: Excel Basics

To activate cells in many rows and columns (i.e., a range) place the cursor in the upper left cell and drag to the lower right. Now all of the cells in that range will appear shaded. You can click and drag in the opposite direction if you wish. This is handier if you tend to overshoot your target and continue on past where you want to end, as many of us do. Another way to activate a range of cells is to click in the upper left cell, move to the lower right cell using the scroll bars or arrows, and then press the SHIFT key and click on the lower right cell (SHIFT + CLICK ON CELL) at the same time. Now all cells between these two points are shaded.

P.5.2 Types of Information

There are three types of information that you can enter into a cell:

1. Text. This is the term used by Excel developers. Other spreadsheet programs may call the alphabetic characters typed in a cell "labels.
2. Numbers. Most often you will enter numerical data.
3. Formulas. These cause new information generated from operations performed on text and numbers that are entered in cells.

P.5.3 Changing Information

To change information in a cell, you have to consider which of two situations exist:

- If you have not yet "accepted" the information by clicking the green check mark (or pressing enter, or using arrow keys, or …) then you can simply use the backspace or delete keys to remove entries. Insert the **I-beam** at the point where you want to change something and use the backspace key to remove characters to its left and the delete key to remove them at its right.

- If you are typing information into a cell and decide you want to start over, click on the red X and everything will be deleted.

- If you just want to delete everything in the **active cell** or range of cells, press the delete button.

- If you have already entered and accepted data in a cell, but now want to go back and edit it but not erase all of it, activate the cell and then insert the I-beam in the editing window where you want to make the changes.

As a practice exercise for changing information, try going through the following steps:

- In cell A1 type 12346. Press Enter, which moves you to A2.
- Assume you really wanted to enter 123456.
- Return to A1 by using the arrow key or mouse.
- Move the mouse pointer so the I-beam is between 4 and 6 in the editing bar. Click once.
- Note that the three editing keys are now shown to the left of 12346.
- Type 5, which will be inserted between 4 and 6

[Spreadsheet screenshot showing cell A1 selected with value 123456, displayed in formula bar and in cell A1 of the grid.]

P.5.4 Moving and copying information

A basic principle used in many programs is that you mark or indicate which material will have something done to it by first marking it and then execute the command that does something to the highlighted material. We will see this principle operate in several other places in the program. When we want to move information we can do it so that it is removed from one location and placed in another. This is a **cut**. If we want to make a copy of the material so that it is in the original location as well as in other locations we make a **copy**. Help for each of these procedures is obtained by searching the Help menu using the words *cutting* or *copying*.

To copy the entry in cell A1 (123456) to cell B1 we first mark the cell by clicking on it. Move the mouse pointer to the icon for copy or use **Edit - Copy**. The outline around the cell will become like the lights on a marquee; it will alternately darken and lighten. It is ready to be copied. Click the cell where you want the copy, say A2. Use the **Edit - Paste** commands or the icon for paste to place a copy there. Copy operates like a rubber stamp; we have a copy stored in the Clipboard and can continue to place copies anywhere we wish. Activate another cell, say C3 and paste another copy there.

Cutting is done in a similar fashion, except that the cell is empty after you cut the data from it. Click on any cell with content, say A1, and then click on the cut icon in the standard toolbar, which is a pair of scissors. The cell will be outlined, as it was when you used *copy*. Activate another cell and click on the icon for pasting, the same one used before.

If you cut or copy a group of cells, i.e., a range, the principle is the same. Instead of activating one cell, you click and drag over the cells so that a range of cells is now marked. Pasting is the same, except that the upper left cell of the group is the one whose location you specify as the target for the cut or copy.

Moving material between worksheets is accomplished the same way. You mark, indicate the cut or copy, move to the new sheet and the location on that sheet, and then paste.

- **Dragging and dropping**. If you want to cut or copy a range of cells on the same sheet, a shortcut to use is called **drag and drop**. Mark the cells, move the cursor to an edge, where it will become an outline arrow, hold down the mouse and drag to the new location. That is for cutting. To copy, hold down the **CTRL** key at the same time. If you attempt to drag and drop material onto a spot that currently has data in it you will get a message that asks if you want to do this, because it will remove the old material. If you have mistakenly done this, you can fall back on the **Undo Drag and Drop** command that you will find in the **Edit** menu. For more information, look up *drag and drop* in the help menu.

P.6 Formatting numbers

Start with a clean spreadsheet. You can clear everything from a sheet by marking the whole sheet (Press CTRL+A) and then pressing the Delete key. Or, you can simply move to another sheet by clicking on a tab at the bottom of the screen. Enter these numbers into A1 through A6 and then copy them into column B, C, D, and E.

	A	B	C	D	E
1	1234567	#######	$1,234,567.00	$1,234,567	123456700%
2	234567	#######	$ 234,567.00	$234,567	23456700%
3	34567	#######	$ 34,567.00	$34,567	3456700%
4	4567	#######	$ 4,567.00	$4,567	456700%
5	567	$ 567.00	$ 567.00	$567	56700%
6	67	$ 67.00	$ 67.00	$67	6700%

- **Currency** Click on the B at the top of the second column to mark it. On the formatting bar you will find a $. Click on it. The initial column width is set at 8.43. With this width, column B looks like the one shown above; all but the last two figures are replaced by ###. When you see this, it indicates that the numbers are too long to fit in that width. We can readjust the column width by double-clicking between the B and C at the top of the column. To show how this will change the appearance, we will activate column C, click on the $ icon and then double-click on the vertical separation between C and D to widen it an appropriate amount. The figure shows what results.

- **Currency variations.** If you wish to change to the format, which has no decimal, click on **Format - Cells**. The dialog box that results indicates the options regarding the number of decimals and the presence or absence of the $. In column E we used this feature to eliminate the cents.

- **Percent and comma** These options are to the right of the **Currency** option. Try them and note the changes. A number like 567 becomes 56700%.

- **Changing the decimal** The two icons to the right of the icon for inserting a comma allow you to increase or decrease the number of places displayed to the right of the decimal. Try these on data you have marked.

P.6.1 Aligning Information

Start with a clearly unaligned set of data, such as the following:

	A	B	C	D
1	1234567			
2	123456.7			
3	12345.67			
4	1234.567			
5	123.4567			
6	12.34567			
7	1.234567			
8	0.123457			

Notice what happens when you enter the last row. It displays 0.123457. The six is missing. The program has been set to display digits to six places to the right of the decimal, if there is space. It rounds off any values that exceed that size. Enter .00001234567. The rounded value of 0.001235 is displayed, but the unrounded number is shown in the edit window. Moving on to .00001234567 you will see the value displayed as 1.23E-05, the exponential form of the number. We will not be concerned with these in this book, but they are commonly used in scientific measurements.

A simple way to align the eight entries above is to mark them and then click on the comma icon. Commas will be placed appropriately and values shown to two decimals. The last entry is displayed as 0.12. As above, the Formula Bar will display what we entered, .1234567, and that is the figure that will be used in computations. You can increase or decrease the number of decimals displayed, as described previously.

P.6.2 Formatting a range

Format a range of cells by marking it and applying the formatting style as was just demonstrated. You can use the Format command on the Menu Bar, followed by the Styles Dialog Box to select the style desired and the number of decimals to display.

P.6.3 Inserting or deleting rows and columns

Inserting or deleting rows and columns is relatively easy. If you have entered data in a row that includes columns A, B, and C and want to place a new column between A and B, do this: Click on the letter B to mark that column. On the Main Menu Bar, select Insert, and then click on **Columns**. You can continue, inserting as many columns as you wish by continuing to click the mouse.

If you are starting over and want to insert two columns between A and B, simply mark columns B and C at the same time and follow the procedure above. To delete these two empty columns, mark them and go to the Edit command and click Delete.

The procedure for inserting or deleting rows is exactly the same, except that you use the numbers for the rows instead of the column letters.

18 Primer: Excel Basics

P.6.4 Filling adjacent cells

Filling can mean one of two things. Filling can mean that you take the contents of a given cell or range of cells and make copies of that material in adjacent cells. Filling, broadly defined, can also mean that you continue a series or sequence of content into adjacent cells. For example, you might want to list the day of the week, and begin by typing Sunday into cell A1 and Monday into cell A2. You then can continue with Tuesday through Sunday in cells A3 through A8 with a simple move. We will discuss this in the section on **Series**.

Filling the same content into adjacent cells:

To learn how to fill the same content into adjacent cells, follow the steps below:

- On a clean spreadsheet, enter Sunday in cell A1 and Monday in cell A2.
- Click in the center of cell A1 and drag to cell A7, activating that range.
- At the menu bar, choose *Edit - Fill* and click on *Down*.
- Now you will have Sunday copied in all the cells from A1 through A7.

Copying takes the contents of the first cell in a series of marked cells and duplicates throughout the marked cells. The content of the second cell, Monday, is removed.
Filling a Series into Adjacent Cells

To learn how to fill a series into adjacent cells, follow the steps presented below:

- As above, enter Sunday in cell A1 and Monday in cell A2.
- Make sure Monday is entered by tapping Enter or clicking on the green check mark.
- Click in the center of cell A1 and drag through A2 so both cells are marked.
- Place the mouse pointer on the black dot (the **fill handle**) at the lower right of the two cells.
- The mouse pointer changes to a solid black plus sign. Drag it to cell A7.

The **series** of days from Sunday through Saturday is listed.

An easier method of accomplishing the same thing is presented below:

- Use the above layout of Sunday in cell A1 and Monday in cell A2
- Click on A1 to activate it.
- Grab the fill handle of cell A1 and drag to A7 to complete the series.

Days of the week are special, since we have a given order determined by convention. If we choose numbers that have no agreed upon sequence, we will not get a series generated by simply choosing the first number on the list.

We can also generate a series of numbers, but the series must have an identifiable pattern.

- On a clean spreadsheet, enter 10 in A1 and 15 in A2.
- Mark A1 and A2 by dragging over them.
- Using the fill handle, drag to A10
- You should have the numbers 10, 15, 20, 25, ... to 55 listed.

P.6.5 Series

The above activities show how easily we can copy the contents of a given cell into adjacent cells or generate a series of numbers from two examples. Now we will use the **series** command to generate a series of numbers.

Follow the steps presented below.

Enter these numbers in cells A1:A3: 5, 7, 9.
Activate the range A1: A10.
Using *Edit - Fill - Series* will produce the dialog box shown below that provides information about the series for you to verify before the program continues the series.

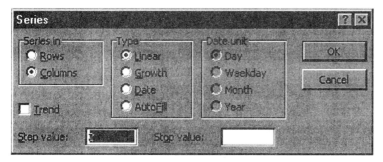

The dialog box displays the program's interpretation of the data. The series is presented in **Columns**. It is a **Linear** relationship. No **Date Units** are provided. The **Step Value** and **Stop Value** are available for verification and determination. If you wanted to specify a value to stop at, it would be entered in that box.

P.6.6 Sorting

Sorting becomes a useful tool when we have a set of data to analyze and want an initial look to see how it is distributed. First mark the data to be sorted and the use the **Data - Sort** commands. There are other ways to sort that we will also describe. Enter the data as shown below into a fresh spreadsheet.

	A	B	C	D
1	Name	Weight	Income	
2	Fred	145	$ 23,000	
3	Sue	123	$ 18,000	
4	John	155	$ 12,000	
5	Jane	118	$ 38,000	
6				

20   Primer: Excel Basics

A dialog box, shown below, requires that we specify which column to use in sorting:

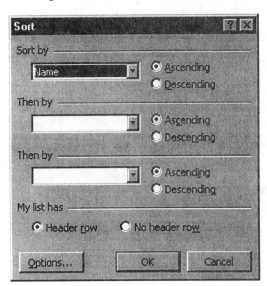

The upper left cell, activated we marked the range, is listed as the default variable to use. We also have the choice of sorting ascending or descending order, using the column. At the bottom of the box we have to indicate if the row at the top is data or titles of the variables. It is correctly marked. If we click OK, the following sort results:

	A	B	C
1	Name	Weight	Income
2	Fred	145	$ 23,000
3	Jane	118	$ 38,000
4	John	155	$ 12,000
5	Sue	123	$ 18,000

Names are now arranged in alphabetical order and the data connected to each name has been moved along with it. Try the above sort, but instead of accepting *Name* as the variable to use in sorting, click on the down arrow, which shows Weight and Income as other choices to use.

Note that we have three options for sorting. In this case we have no data that is the same for a variable, e.g. no two people weigh the same. If, for example, we had a list of 100 student members of an organization that included active and former membership status and address (street, city, zip code), then we would have a number of zip codes that are identical. In the case we might want to sort by zip code, and then by membership status if we were generating mailing labels (sorted by zip code for a cheaper rate) and different messages for different types of members. The one identifier that should not have any duplicates is social security account number, which is why it is often used as an identifier.

P.7 Saving and Retrieving Information

P.7.1 Naming workbooks

Earlier versions of Excel used one sheet as the basic unit that was stored as one file. Now, up to 256 sheets can be stored as one unit, called a workbook. If you have opened a new workbook and entered data that you want to save, the logical next step is to click on File in the Main Menu and click on *Save*. When you have a new, previously unnamed sheet, you're then prompted to give the file a name. The default name of Book 1 appears in the File Name window of the Save As dialog box. Unless you indicate otherwise, the file will be saved in the last location used, be it the hard drive, network server, or diskette. A common default is to save in a folder called *My Documents*, which is found by clicking on the file folder icon having an asterisk (*) in it. This is called *Favorites*. The following figure shows the *Save As* dialog box:

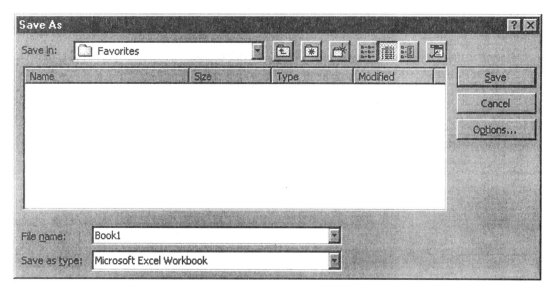

In version 7.0 of Excel, using Windows 95 and subsequent versions, files can have names longer that eight letter. In earlier versions, the **DOS** naming convention applies. In that case, a file cannot have a name longer than eight letters, followed by a period and an extension, usually .xls. A typical file name might be Bus101hw.xls. There are restrictions on the symbols that you can use in a file name, which will be apparent when you get an error message for using a/or some other forbidden symbols. Windows 95 and subsequent versions, and earlier Mac versions allow names to have up to 256 characters, so they can be more descriptive of the contents. Use the *?* and *Help* to learn the names and functions of the various boxes in the *Save As* dialog box.

P.8 Printing

P.8.1 Page Setup

The **Page Setup** dialog box displays four tabs along the top to access the four options: **Page, Margins, Header/Footer,** and **Sheet**. It is accessed by going to File - Page Setup.

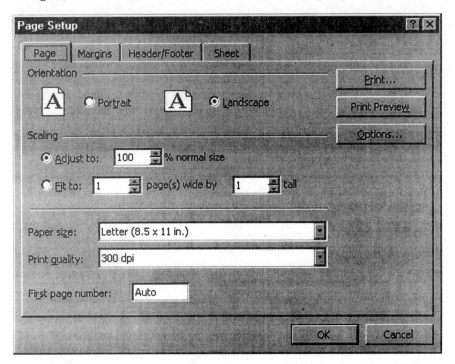

P.8.2 Page

Page allows the user the change the **orientation** of the page from portrait (the way this book is printed) to landscape (sideways). With the **scaling** commands you can adjust the size of the image to be 10% of normal size or enlarge it to 400%. The **Fit to** command instructs the program to automatically adjust the size to fit any of a number of pages as specified.

> Beware: If you have three or four pages of material to print and direct the program to fit it to one page, it is sometimes difficult to read. You can specify that it will stay one page high and allow it to continue to the right on other sheets as far as necessary by using the two adjustments available.

Paper Size and **Print Quality** provide options shown via the drop-down arrow, although most often letter size paper is used, and print quality adjustments are not available. On some printers, you can move to a higher quality print by changing the dpi to a value of 600 or more. This does increase printing time and use of ink.

Clicking on **Options** provides more graphic views of some of these choices, depending on your printer.

First Page Number is set to **Auto** as the default. If you wish to start the page numbering at a specific value, type in that value after highlighting **Auto**.

P.8.3 Margins

Margins are set using the options shown by clicking the Margins tab. The preview window shows how these will look as you change them. Note that you can also center the output on the page horizontally, vertically, or both. After you have changed any of the settings you can click on **Print Preview...** to check its appearance. You can also change the margins in Print Preview by dragging the margin boundary handles.

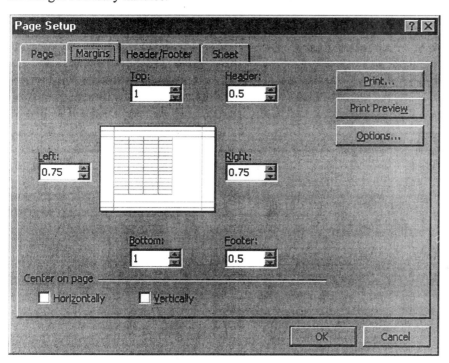

P.8.4 Header/Footer

Headers contain the information printed across the top of all (or all but the first) pages. In this text the headers on one page indicate the chapter title and page number; on the other we have the section and page number. Sometimes this information is printed at the bottom of a page, and called a **Footer**. Our descriptions of the procedures for developing headers also applies to footers.

24 Primer: Excel Basics

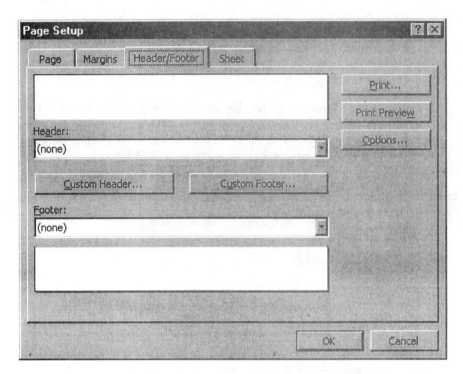

Unless you alter the settings, the **header** and **footer** shown above are the default. Click on the arrow to the right of the default header and you will see some previously used on the computer. You may wish to choose one of them and edit it to your liking. Or, if you wish, you can click on **Options...** and obtain the following dialog box:

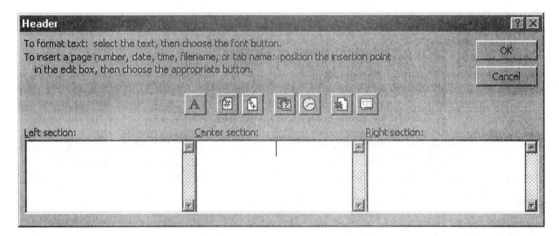

In this case, you have the option to place your header in the left, center, or right of the page. Use the ? and the help to determine what the icons will do for you. The letter A, when clicked, allows you to write in text, changing the font and size; others are for page numbering, date and time, filename and sheetname.

P.8.5 Sheet

This page allows you to choose the area to print, whether to print titles of columns or rows on sheets that continue on to other sheets, print gridlines, the quality of print, and row and column headings. Click on the ? and point to each of the areas to learn how to use them.

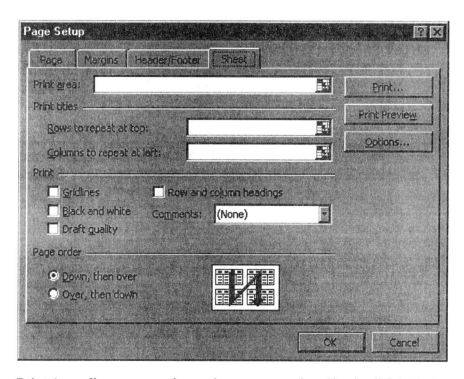

Print Area allows you to select a given range to print. Simply click in the box to the right and then move to the spreadsheet where you drag to indicate the range to select. On Version 7 you will have to grab the above window by the title bar to move it away from the area you want to mark. Version 8 has an option for briefly collapsing this dialog box to more easily do this.

P.8.6 Enhancing output

With a bit of experience you can improve the appearance of your output by using the options described above. Using the preview option allows you to see exactly how things will look before you print them. Sometimes gridlines improve the display. An enlarged title adds clarity. Size of the sheet can be adjusted for your purposes. All of these can be explored using the options above.

P.8.7 Inserting and removing page breaks

When a worksheet spans more than one page, Excel determines where to insert a page break. You might find it more convenient to make the break at a different location. Use the Help function to look up *page breaks* in the index for directions. If you have use *Page* in Page Setup to fit everything onto one page, it may not appear as you wish, especially if the material takes up less than one page. Page Preview is always a good way to determine what looks best.

P.8.8 Preview and print

Clever computer users preview their material before they print it. You can go to the Print Preview command under File or simply click the icon on the Standard Toolbar. To change options, after you view your sample, you can use the menu at the top. Try out such things as Zoom, which will increase the size of the page so you can view details. Click it to turn it on and click it to return to the full page view. Setup will display the four pages described above, in that section. Margins will display lines on your preview that shows where the margins will be located. You can grab the handles and change them on the sheet. Close returns you to the regular view of the page. Print moves you to the dialog box with choices regarding the printer to use, number of copies, which pages to print, etc.

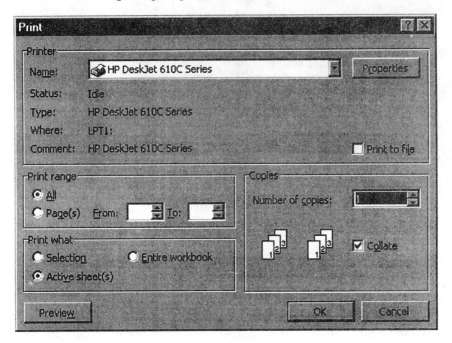

P.9 Using Formulas and Functions

P.9.1 Operators

Algebraic formulas utilize the four common operations of addition (+), subtraction (−), multiplication (×), and division (÷). When creating formulas, we will use these operations along with three others. Called **operators**, they are symbolized as:

    Addition +   Multiplication *   Subtraction −   Division /

Negation refers to using the minus sign to indicate a negative number, as in −3

    Exponentiation ^   Percent %

Multiplication uses an asterisk instead of an ×. Division uses a diagonal (/) instead of the division symbol. If we wish to raise a value to a power, say X^2 we palce the carat (^) between the X and the 2. To raise a number X to the second power, we would write X^2. Finally, we can also place a percent (%) sign behind a value, as in 20%. For example, the formuls 15 ^ 2 * 15% raises 15 to the second power and multiplies the result by 0.15 (the decimal form of 15%) to produce the result of 33.75.

We also group sets of operations together with parentheses, which determines the order of execution of commands by the computer. Any operations enclosed in parentheses are executed first, moving from the innermost parentheses to the outermost.

P.9.2 Order of Operations

The order of operations in Excel is:

1. Negation, as in –15
2. Percent
3. Exponentiation
4. Multiplication and division
5. Addition and subtraction

Excel first calculates expressions in parentheses and then uses those results to complete the calculation of the formula. For example

= 2+4*3	produces 14 because multiplication occurs before addition
=(2+4)*3	produces 18 because operations within parentheses are executed first
=2+(4*3)	produces 14 because operations within parentheses are executed first

In a more complex situation,

=2+(4*(3+5)^2)/2   produces 130.

First the terms in the innermost parentheses are executed, producing 8. Then this is raised to the second power, producing 64, since this is within the second set of parentheses and exponentiation has precedence over the other operations. 64 is multiplied by 4, yielding 256. This is divided by 2, producing 128. Finally added to 2 yields 130. Try this on your computer, by entering this formula into cell A1 and pressing **Enter**.

P.9.3 Writing Equations

In the equations above, we have always used numeric constants, such as 2, 3, 4, or 5. In practice, the formula that you write will probably refer to a cell that may contain any value. This is equivalent to the X and Y that acted as unknowns in algebra. For example, to convert degrees Fahrenheit to Celsius, we use the formula:

C = 5/9(F-32)

Let's set up a table to provide this conversion. Type the label in cell A1: *Degrees Fahrenheit*. In cell B1 type: *Degrees Celsius*. Enter the two values of –60 and –59 in cells A2 and A3. That's a good place to start if you live in Wisconsin in January.

	A	B
1	Degrees Farenheit	Degrees Celsius
2		
3	-60	
4	-59	
5		
6		

28 Primer: Excel Basics

Let Excel continue the series by clicking on the value of –60 and dragging down to cell A80 to mark those cells. Then go to Edit – Fill – Series and notice that the program has correctly inferred that the step size is +1. Each time we move down a row the value in column A is reduced by one unit. We wish to continue the series with that change. Click OK and we have data extending from –60 to 17 degrees. In cell B3 enter this formula:

=5/9(A3-32)   and press Enter.

When the temperature is –60 degrees Fahrenheit, it is –51.11 degrees Celsius, as cell B3 indicates. Activate cell B3. Now, just click on the small square in the lower right of cell B3 (the fill handle), and drag to cell B80. The screen pointer becomes a solid black plus sign when you use the fill handle. You have copied the formula into all of those cells, with the address of each cell that represents degrees Fahrenheit changing for each row. You should have something like the following:

	A	B
1	Degrees Farenheit	Degrees Celsius
2		
3	-60	-51.11111111
4	-59	-50.55555556
5	-58	-50
6	-57	-49.44444444
7	-56	-48.88888889
8	-55	-48.33333333
9	-54	-47.77777778
10	-53	-47.22222222
11	-52	-46.66666667
12	-51	-46.11111111
13	-50	-45.55555556
14	-49	-45
15		

To make our output more attractive, click on column B to activate that column, go to Format – Cells – Number and accept the default of two places to the right of the decimal. Now you should have the first few rows of data looking like this:

	A	B
1	Degrees Farenheit	Degrees Celsius
2		
3	-60	-51.11
4	-59	-50.56
5	-58	-50.00
6	-57	-49.44
7	-56	-48.89
8	-55	-48.33
9	-54	-47.78
10	-53	-47.22
11	-52	-46.67
12	-51	-46.11
13	-50	-45.56
14	-49	-45.00
15		

P.10 Entering Formulas

Now, let's take a situation that is very commonly used by students. Compute a Grade Point Average. Enter the following information in a new blank spreadsheet.

	A	B	C	D	E
1	Course	Credits:	Grade:	Value:	
2					
3	English	5	A	4	
4	Math	4	B	3	
5	Psychology	3	A	4	
6	Economics	3	C	2	
7		15.00			
8					
9	Grade:	Value:			
10	F	0			
11	D	1			
12	C	2			
13	B	3			
14	A	4			
15		5			

We want to compute the Grade Point Average. We have to multiply the credits for a course by the value of the grade given for that course. In Cell E3 enter this formula: =B3*D3. Then use the fill handle in the cell to copy the formula from E3 to E6, so that you have the product of credits times value of grade for each course. The recommended procedure for setting up formulas is to click on the cell whose address you want entered, NOT type in the address of that cell. In this case you would type the = sign in E3 and then click on B3, which now appears in the formula. Then enter an asterisk (from either SHIFT-8 or the * key on the keypad) followed by clicking on D3. Then either click on Enter or the green arrow in the formula bar. You should have:

	A	B	C	D	E
1	Course	Credits:	Grade:	Value:	Grade Points:
2					
3	English	5	A	4	20
4	Math	4	B	3	12
5	Psychology	3	A	4	12
6	Economics	3	C	2	6
7		15.00			
8					
9	Grade:	Value:			
10	F	0			
11	D	1			
12	C	2			
13	B	3			
14	A	4			
15		5			

Now all we need to do is to sum the Grade Points, sum the Credits, and then divide Grade Points by Credits to get GPA. Use the AutoSum button, Σ (a Greek Sigma on the Standard Toolbar) to obtain the sum for each. Just activate cell B7 and click on that button. A series of dotted lines surrounds the four digits above, indicating that the program estimates that this is what you want to sum. Click Enter and the value of 15 will be displayed. Do the same for the Grade Points to obtain 50. To obtain the GPA you can enter those two values in an equation directly or enter the cell references in the equation to obtain the answer. If you do the latter, your equation would look like: =E7/B7. Put your answer in cell E10 as shown below:

	A	B	C	D	E
1	Course	Credits:	Grade:	Value:	Grade Points:
2					
3	English	5	A	4	20
4	Math	4	B	3	12
5	Psychology	3	A	4	12
6	Economics	3	C	2	6
7		15.00			
8					
9	Grade:	Value:		GPA=	3.333333333
10	F	0			
11	D	1			
12	C	2			
13	B	3			
14	A	4			
15		5			

Note that the value of 3.3333… doesn't appear until you press Enter of click on the green arrow. If you click on cell E9 now, the formula you used to obtain that value will be displayed in the Formula Bar, not the value shown here. For emphasis, you can mark cells D9 and E9 and then click on the symbol to make them **BOLD**.

P.10.1 Relative References

When we entered the formula for determining Grade Points in cell E3, we entered it using the relative reference mode. We told the program to go to cell B3, obtain that value, go the cell D3, obtain that value, multiply them and place the answer in E3. Actually we instructed Excel to move the three cell locations to the left (B3) of the active cell, obtain that number, move one cell to the left (D3) of the active cell, obtain that number, multiply them and place the answer in the active cell. To see how this works, try this: Activate cell E3. Note the formula in the Formula Bar. Now activate cell F3. The formula references in the formula bar have been adjusted to reflect what was said above; the program is not going to absolute locations of cells, but is moving relative to the active cell, using the same movements in the spreadsheet that were specified in the formula. Now let's make our computation of the GPA a bit more sophisticated by using that table that gives the value of each grade. We will use a function called **table lookup**.

P.10.2 Absolute References

When we have a table that converts a letter grade to a corresponding number (or vice versa), we can use a built-in function known as **VLOOKUP**. Mark cells D3 through D6 and press delete to clear the contents. We will have the program **VLOOKUP** compare the grade letter printed in cells C3 through C6 to the tabled values in cells A10:B14 to fill in the number value corresponding to a letter grade. Click on the Function Wizard (the f_X on the Standard **Toolbar**). Move to the **Lookup & Reference** part of the Paste Function Window as shown below. Click on **VLOOKUP** as indicated. Then click OK.

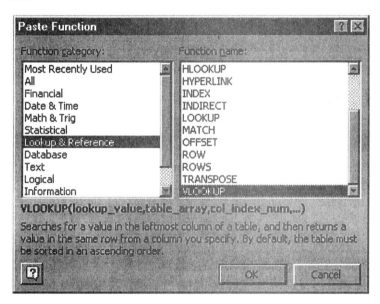

The following window appears after you execute the above and enter the values into the formula as shown in the formula bar. After we enter the =VLOOKUP in cell D3 we enter the location of the grade to translate (cell C3), then the absolute address of the table ($A$10:$B$14), the letter 2 to indicate that we want the number from the second column in the table, followed by FALSE which indicates that we want an exact match. After accepting the formula by clicking the green arrow, we use the fill handle to copy the formula into cells D4: D6. The cells $A$10:$B$14 are absolute references because we do not want them to change as we refer to them in the operations on the second, third, and fourth grades. Whenever we want an address to be absolute, we preface the usual column letter and row number with a dollar sign. The figure below shows the formula entered into cell D3 and the resulting value of 4. Note how the C3 value is a relative value while references to the table are absolute.

32 Primer: Excel Basics

D3		=	=VLOOKUP(C3,A10:B14,2,FALSE)	
	A	B	C	D	E
1	Course	Credits:	Grade:	Value:	Grade Points:
2					
3	English	5	A	4	20
4	Math	4	B	3	12
5	Psychology	3	A	4	12
6	Economics	3	C	2	6
7		15.00			
8					
9	Grade:	Value:		GPA=	3.333333333
10	F	0			
11	D	1			
12	C	2			
13	B	3			
14	A	4			
15		5			

Chapter 1
Statistics, Data, and Statistical Thinking

1.1 Introduction

Chapter 1 introduction the topics that will be expanded on throughout the text. No data analysis is necessary in Chapter 1 and Microsoft Excel® cannot be used here in the text.

Chapter 2
Methods for Describing Sets of Data

2.1 Introduction

Chapter 1 served to introduce many of the basic statistical concepts employed in all types of data analysis problems. Two main areas of statistics emerge from Chapter 1 - descriptive and inferential statistics. Chapter 2 focuses on the descriptive area and looks at both graphical and numerical techniques that allow statisticians to summarize data that has been collected. Many of the techniques used to summarize data discussed in *Statistics for Business and Economics* can easily be performed with Excel. Our purpose is to explain these techniques and to illustrate them using the examples presented in the text as well as additional examples provided here. Listed below are the various techniques that Excel offers that can be used to generate the graphical and numerical topics presented in Chapter 2.

Excel offers a wide array of graphing options to the statistician. When working with qualitative data, Excel allows the statistician to create customized **pie charts** and **bar graphs**. For quantitative data, **stem-and-leaf displays, boxplots, histograms,** and **scatter plots** are easy to create. The scatter plot feature in Excel can readily be used to create the **time series plot** discussed in the text. At the present time, Excel does not allow the creation of **dot plots**.

As with most database and statistical software programs, Excel provides a wide array of numerical description of data. The three measures of central tendency (**mean, median,** and **mode**) and the three measures of variability (**range, variance,** and **standard deviation**) are all available in the descriptive statistics menu of Excel. Measures of relative standing (**percentiles, quartiles,** and **z-scores**) are available in Excel but not as easy to access as the measures of central tendency and measures of spread. The **box plots** discussed in Chapter 2 are not available in Excel.

The following examples from *Statistics for Business and Economics* are solved with Microsoft Excel® in this chapter:

Excel Companion		Statistics for Business and Economics		Data Set
Example	Page	SBE	SBE Page(s)	
2.1	36	Example 2.2	41-44	SBE Example 2-2
2.2	42	Table 2.1	29	SBE Table 2-1
2.3	44	Example 2.18	97-98	SBE Example 2-18
2.4	49	Example 2.2	41-44	SBE Example 2-2
2.5	50	Example 2.2	41-44	SBE Table 2-3
2.6	51	Example 2.6	57	SBE Table 2-3
2.7	54	Example 2.9	69	SBE Table 2-3
2.8	55	Example 2.9	69	SBE Table 2-3
2.9	56	Example 2.9	69	SBE Table 2-3

2.2 Graphical Techniques in Excel

2.2.1 Bar Graphs and Histograms

Bar graphs, pie charts, and scatter plots are all easy to generate using Excel. The graphs enable the user summarize the data that they are viewing and make decisions quickly and easily. **PHStat** offers an easy method of creating pie bar graphs for qualitative data within the **One-Way Tables and Charts** menu. For quantitative bar graphs, we will utilize the **Histogram Data Analysis** procedure within the Excel program.

Example 2.1: As an example we turn to Example 2.2 from *Statistics for Business and Economics* found on pages 41-44 of the text:

A manufacturer of industrial wheels suspects that profitable orders are being lost because of the long time the firm takes to develop price quotes for potential customers. To investigate this possibility, 50 requests for price quotes were randomly selected from the set of all quotes made last year, and the processing time was determined for each quote. The processing times are displayed below in Table 2.1, and each quote was classified according to whether the order was "lost" or not (i.e., whether or not the customer placed an order after receiving a price quote).

Table 2.1

Request	Time	Lost?	Request	Time	Lost?
1	2.36	No	26	3.34	No
2	5.73	No	27	6.00	No
3	6.60	No	28	5.92	No
4	10.05	Yes	29	7.28	Yes
5	5.13	No	30	1.25	No
6	1.88	No	31	4.01	No
7	2.52	No	32	7.59	No
8	2.00	No	33	13.42	Yes
9	4.69	No	34	3.24	No
10	1.91	No	35	3.37	No
11	6.75	Yes	36	14.06	Yes
12	3.92	No	37	5.10	No
13	3.46	No	38	6.44	No
14	2.64	No	39	7.76	No
15	3.63	No	40	4.40	No
16	3.44	No	41	5.48	No
17	9.49	Yes	42	7.51	No
18	4.90	No	43	6.18	No
19	7.45	No	44	8.22	Yes
20	20.23	Yes	45	4.37	No
21	3.91	No	46	2.93	No
22	1.70	No	47	9.95	Yes
23	16.29	Yes	48	4.46	No
24	5.52	No	49	14.32	Yes
25	1.44	No	50	9.01	No

a. Use a statistical software package to create a frequency histogram for these data. Then shade the area under the histogram that corresponds to lost orders.

b. Use a statistical software package to create a stem-and-leaf display for these data. Then shade each leaf that corresponds to a lost order.

c. Compare and interpret the two graphical displays of these data.

We answer part a. by utilizing the bar graph utility within the **PHStat** program. Before we begin, we must access the data set for this example. **Open** the Data File **SBE Example 2.2** which is found on the floppy disk included with this manual. If done correctly, the data should appear in a workbook similar to that shown below in Figure 2.1

Figure 2.1

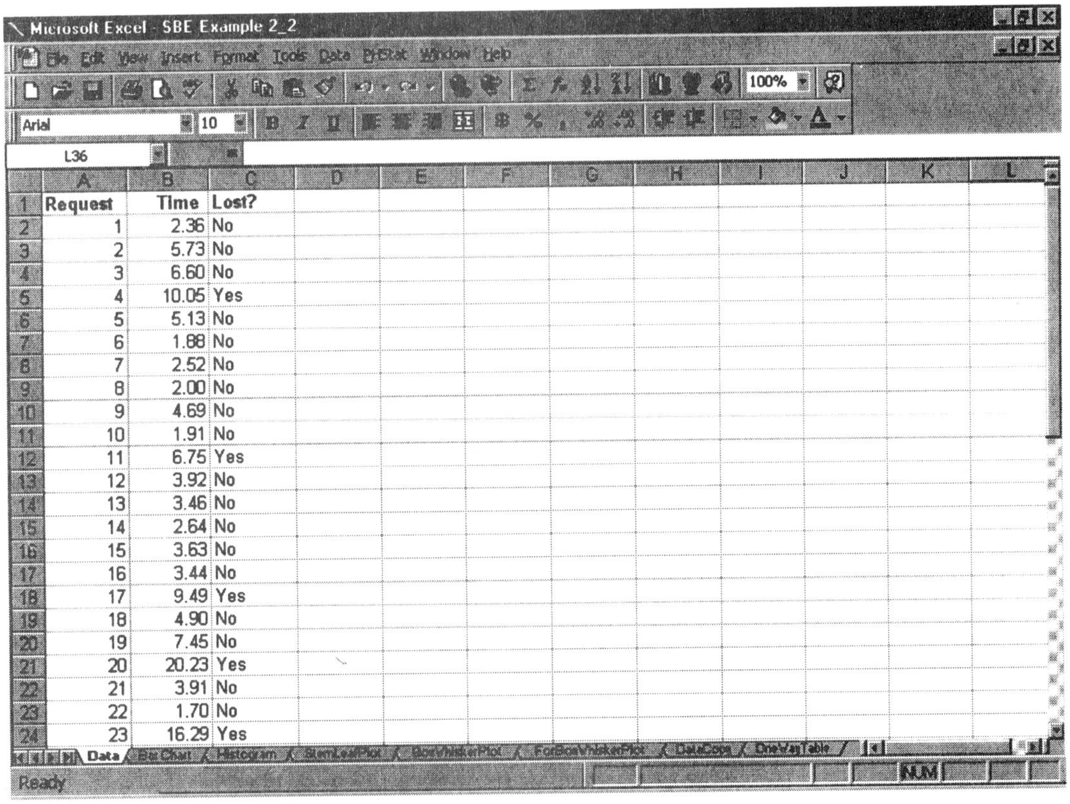

Request	Time	Lost?
1	2.36	No
2	5.73	No
3	6.60	No
4	10.05	Yes
5	5.13	No
6	1.88	No
7	2.52	No
8	2.00	No
9	4.69	No
10	1.91	No
11	6.75	Yes
12	3.92	No
13	3.46	No
14	2.64	No
15	3.63	No
16	3.44	No
17	9.49	Yes
18	4.90	No
19	7.45	No
20	20.23	Yes
21	3.91	No
22	1.70	No
23	16.29	Yes

Our goal in part a. is to create a bar graph for the **Lost?** variable in the data set. Choose the **One-Way Tables and Charts** option within the **PHStat** menu. Begin by entering the rows and columns where the data is located in the **Variable Cell Range** of this menu (see Figure 2.2). This can be done by **typing** the location or by **clicking and dragging** over the appropriate data cells in your workbook. You have the option of including the variable name in the first cell of data. If selected, the variable name will appear on the graph constructed. Lastly, we must **check** the type of chart we wish to construct. In this example, we select the **Bar Chart** option. Click **OK**.

Figure 2.2

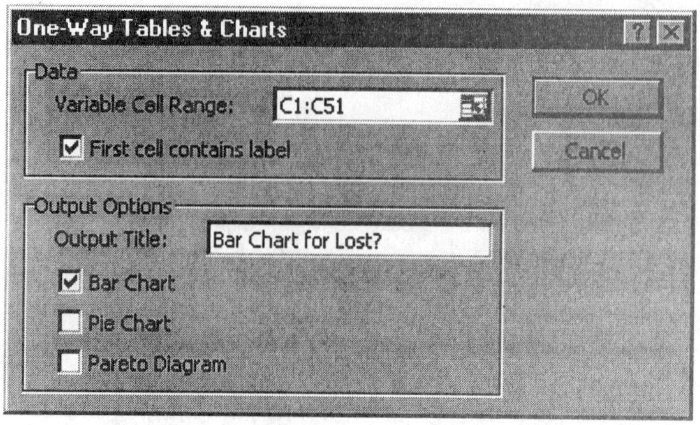

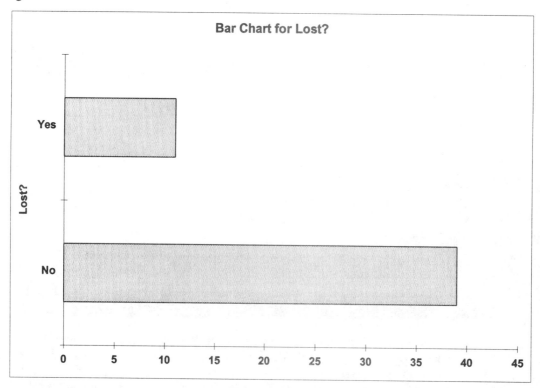

The bar chart generated by this menu is shown below in Figure 2.3. Note that some chart options were changed to display the bar chart in a black-and-white format.

Figure 2.3

To generate a histogram for the quantitative data of this example we utilize the histogram technique found within the Excel program.

Once the data is available for analysis, select the **Tools** icon from the top of the Excel worksheet. Choose the Data Analysis option from within the Tools menu. From the Data menu choose the Histogram option (see Figure 2.4.) Click OK.

Figure 2.4

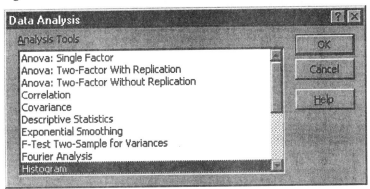

Perform these steps to enter the histogram menu within Excel. There are many options available within this menu. Our purpose here is to demonstrate the easiest method necessary to generate a histogram of the data, and to show the commands necessary to generate a histogram that matches the one shown in the text.

The easiest method to generate a histogram is shown in Figure 2.5. First, enter the rows and columns where the data is located in the **Input Range** of the histogram menu. This can be done by **typing** the location or by **clicking and dragging** over the appropriate data cells in your worksheet. The next step is to specify the **Output Range**. We have chosen to begin the output at column F, row 1 by **typing F1** in the Output Range line of the histogram menu. We have the option in Excel 97 to place this output in a new worksheet by specifying the New Worksheet Ply option. Finally, in order to generate the histogram, it is necessary to **check** the **Chart Output** option in the histogram menu. Click **OK**.

Figure 2.5

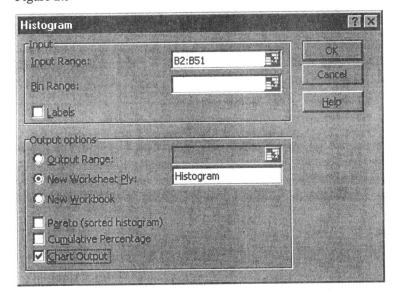

40 Chapter 2: Methods for Describing Sets of Data

Excel generates two pieces of information and places this output beginning at the location we specified above. The first is a table of the data that is being charted. The table contains two pieces of information, Bin and Frequency. Bin (see Table 2.2) refers to the upper endpoint of the histogram bar that is to be drawn and Frequency is the number of observations that will be included in the corresponding bar.

Table 2.2

Bin	Frequency
1.25	1
3.961428571	17
6.672857143	16
9.384285714	8
12.09571429	3
14.80714286	3
17.51857143	1
More	1

Together, this information is used by Excel to generate the histogram (see Figure 2.6). The size of the histogram can be altered to make viewing the chart easier. Simply click on the histogram and stretch the squares on the outline of the histogram to make the display larger or smaller.

Figure 2.6

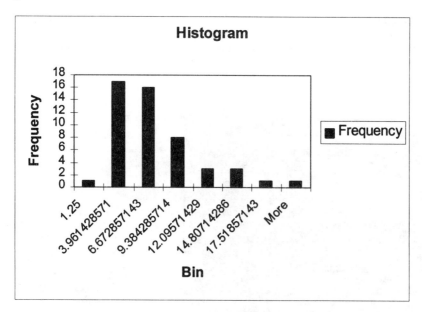

These commands generate a histogram that summarizes the data. Comparing this histogram to the one shown in the text, we see two major differences. First, this histogram has displayed the bars as being separated from one another, while the histogram in the text has bars that touch. The second difference is that the intervals, or interval endpoints, used in the two histograms differ. Both of these differences can be addressed using various options within Excel.

It is important to emphasize that no one histogram that is produced from the data is considered the "correct" one. Our purpose in duplicating the histogram presented in the text is to introduce the user to some of the many options that are available within the histogram menus of Excel. Producing a histogram comparable to the one in the book will allow for easier comparison with the stem-and-leaf display and for easier interpretation of the results.

The touching bars can be adjusted by **clicking** on any one of the bars generated in the histogram above. Next, select the **Format** option listed at the top of the Excel worksheet. Click on the **Selected Data Series** option. Select the **Options** tab of the Format Data Series menu (see Figure 2.7). The option that will allow the bars to touch is the **Gap Width** selection. Change the Gap Width to 0 to assign no gap between bars.

Figure 2.7

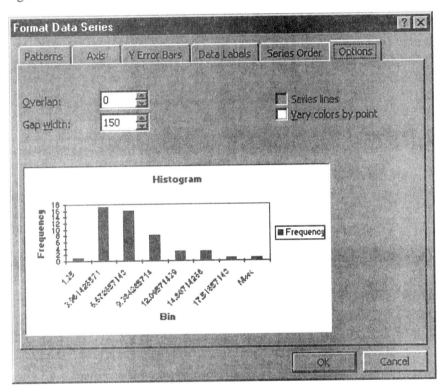

To change the intervals used by Excel requires the addition of a new column in the dataset used for the analysis. You must specify what Bin values that Excel should use to construct the histogram. The Bin values represent the largest endpoint of the bars generated in the chart. To duplicate the histogram presented in the text, with intervals of 2 days, the bar endpoints should be chosen at 3, 5, 7, 9, 11, 13, 15, 17, 18, and 21. This column of values must be entered alongside the data and chosen as the Bin Range (see Figures 2.8 and 2.9).

42   Chapter 2: Methods for Describing Sets of Data

Figure 2.8

Bin Values
3
5
7
9
11
13
15
17
18
21

Figure 2.9

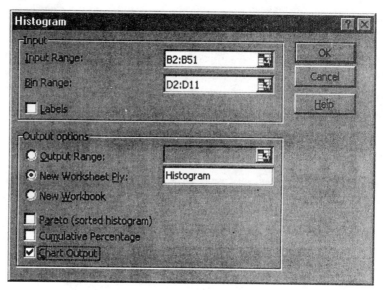

Together, these two changes can be used to produce a histogram that is like the one shown in the text (see Figure 2.10). Be sure to read through Example 2.2 in the text to understand how to interpret these results.

Figure 2.10

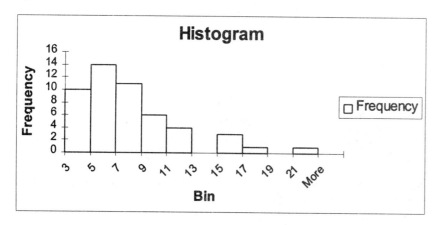

2.2.2 Pie Charts

The second type of graphical technique constructed by Excel is the pie chart. Since none of the chapter examples from *Statistics for Business and Economics* specifically ask for a pie chart, we will use the data from Table 2.1 found of page 29 and create the corresponding pie chart.

Example 2.2:   **Open** the Data File **SBE Table 2.1,** which is found on the floppy disk included with this manual. If done correctly, the data should appear in a workbook similar to that shown below in Figure 2.11.

Figure 2.11

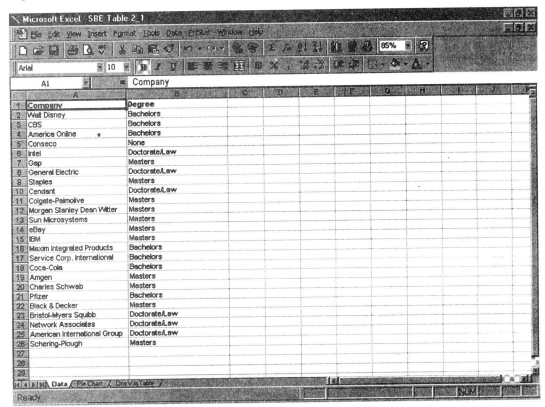

Our goal in part a. is to create a pie chart for the **Degree** variable in the data set. Choose **the One-Way Tables and Charts** option within the **PHStat** menu. Begin by entering the rows and columns where the data is located in the **Variable Cell Range** of this menu (see Figure 2.12). This can be done by **typing** the location or by **clicking and dragging** over the appropriate data cells in your workbook. You have the option of including the variable name in the first cell of data. If selected, the variable name will appear on the pie chart constructed. Lastly, we must **check** the type of chart we wish to construct. In this example, we select the **Pie Chart** option. Click **OK**.

Figure 2.12

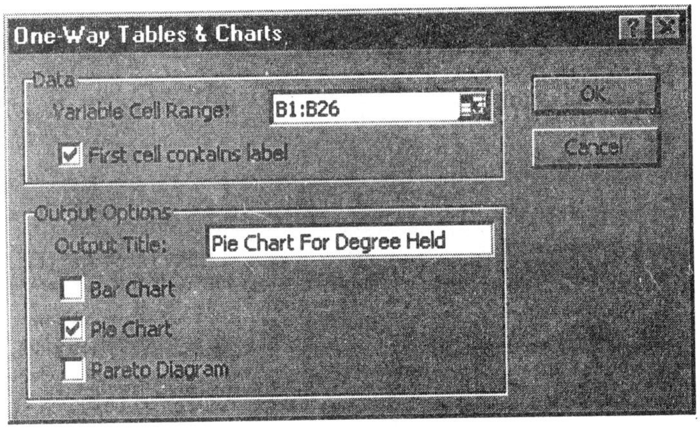

44 Chapter 2: Methods for Describing Sets of Data

The pie chart generated by this menu is shown below in Figure 2.13. Note that some chart options were changed to display the bar chart in a black-and-white format.

Figure 2.13

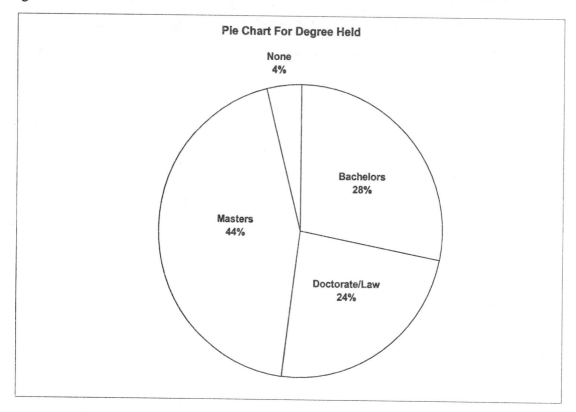

2.2.3 Scatter Plots

The final graphing technique discussed in *Statistics for Business and Economics* that can be constructed within Excel is the scatter plot. We use Example 2.18 from the text to demonstrate how to conduct a scatter plot.

Example 2.3: *Statistics for Business and Economics* Example 2.18 found on pages 97-98.

A medical item used to administer to a hospital patient is called a **factor**. For example, factors can be intravenous (IV) tubing, IV fluid, needles, shave kits, bedpans, diapers, dressings, medications, and even code carts. The coronary care unit at Bayonet Point Hospital (St. Petersburg, Florida) recently investigated the relationship between the number of factors administered per patient and the patient's length of stay (in days). Data on these two variables for a sample of 50 coronary care patients are given in Table 2.3. Use a scattergram to describe the relationship between the two variables of interest, number of factors, and length of stay.

Table 2.3

Number of Factors	Length of Stay (in Days)	Number of Factors	Length of Stay (in Days)	Number of Factors	Length of Stay (in Days)
231	9	233	8	115	4
323	7	260	4	202	6
113	8	224	7	206	5
208	5	472	12	360	6
162	4	220	8	84	3
117	4	383	6	331	9
159	6	301	9	302	7
169	9	262	7	60	2
55	6	354	11	110	2
77	3	142	7	131	5
103	4	286	9	364	4
147	6	341	10	180	7
230	6	201	5	134	6
78	3	158	11	401	15
525	9	243	6	155	4
121	7	156	6	338	8
248	5	184	7		

Solution:

Excel offers a **Chart Wizard** to help create a variety of charts. It is easy to use and we now will demonstrate with the data from the example above. **Open** Data File Example 2.18. To enter the Chart Wizard, Click on the **Insert** menu located at the top of the Excel worksheet. Click the **chart** option within the Insert menu. **Highlight** the XY(Scatter) option within the Step 1 - Chart Type of the Chart Wizard (See Figure 2.14) **Click Next** to advance to Step 2 - Chart Source Data. Both variables to be plotted need to be included in the **Data** range entry within this menu. Make sure that the first row or column of data pertains to the variable to be plotted on the x-axis. The second row column of data should pertain to the y-axis variable. Enter this **Data Range** (see Figure 2.15) and **click Next**. Step 3 - Chart Options allows the user to specify many graphing options. We only mention that it is here that titles can be added to the scatter plot to ease the understanding of the graph (see Figure 2.16). We leave the user to experiment with the other options available at this step. **Clicking Next** allows the user to finish the Chart Wizard in Step 4 - Chart Location. The user may specify where the constructed scatter plot will appear in Excel (see Figure 2.17). Fill in the location as either a new worksheet or as an object and **click Finish**. The finished plot for this example appears in Figure 2.18.

Figure 2.14

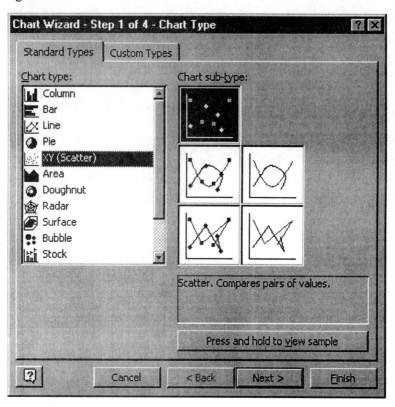

Section 2.2: Graphical Techniques in Excel 47

Figure 2.15

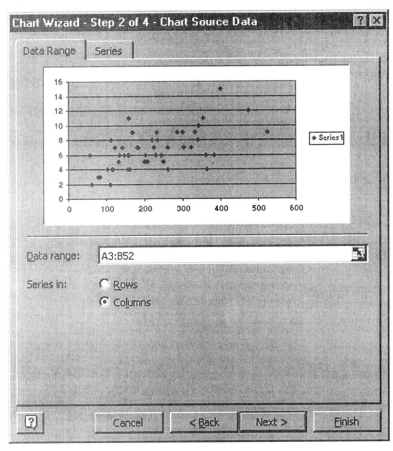

Figure 2.16

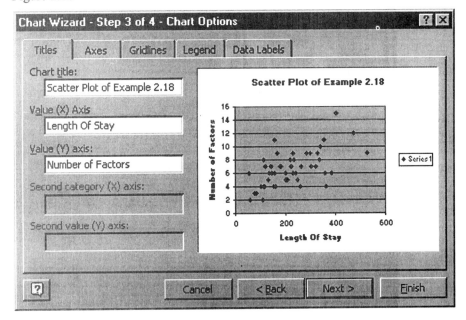

48 Chapter 2: Methods for Describing Sets of Data

Figure 2.17

Figure 2.18

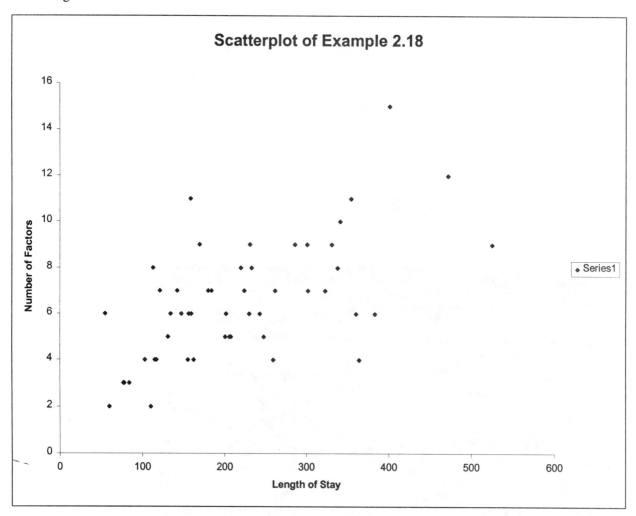

Section 2.2: Graphical Techniques in Excel 49

The scatter plot can also be used to create the time series plot discussed in *Statistics for Business and Economics*. When creating the scatter plot, the measure of time over which the data was collected should be used as the x-axis variable in the time series plot. All other steps are identical to that of the scatter plot discussed here.

2.2.4 Stem-and-Leaf Displays

PHStat allow the user to easily create stem-and-leaf displays for quantitative data. We demonstrate by using Example 2.2 part b., found on pages 41-44 of the *Statistics for Business* and Economics text.

Example 2.4: Use PHStat to create a stem-and-leaf display for these data.

Solution: We answer part b. by utilizing the Stem-and-Leaf Display utility within the **PHStat** program. Before we begin, we must access the data set for this example. **Open** the Data File **Example 2_2** which is found on the floppy disk included with this manual. Choose the **Stem-and-Leaf Display** option within the **PHStat** menu. Begin by entering the column where the data is located in the **Variable Cell Range** of this menu (see Figure 2.19). This can be done by **typing** the location or by **clicking and dragging** over the appropriate data cells in your workbook. You have the option of including the variable name in the first cell of data. If selected, the variable name will appear on the graph constructed. You also have the option of specifying a title for the stem-and-leaf display constructed. Click **OK**. The resulting stem-and-leaf display is shown in Figure 2.20.

Figure 2.19

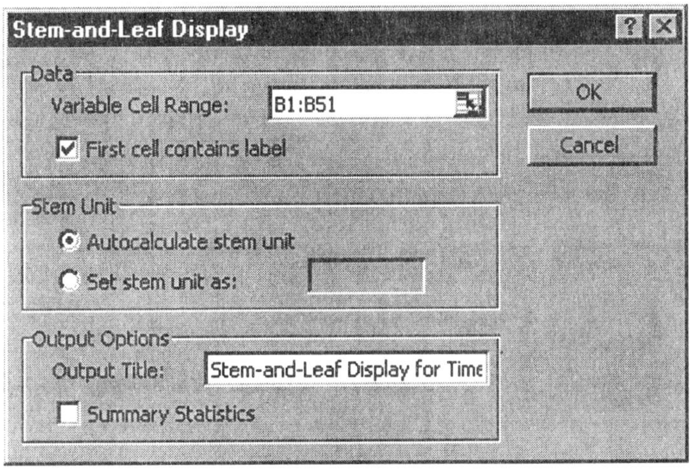

Figure 2.20

```
Stem-and-Leaf Display
for Time
Stem unit:    1

     1|3 4 7 9 9
     2|0 4 5 6 9
     3|2 3 4 4 5 6 9 9
     4|0 4 4 5 7 9
     5|1 1 5 5 7 9
     6|0 2 4 6 8
     7|3 5 5 6 8
     8|2
     9|0 5 9
    10|1
    11|
    12|
    13|4
    14|1 3
    15|
    16|3
    17|
    18|
    19|
    20|2
```

50 Chapter 2: Methods for Describing Sets of Data

2.2.5 Box Plots

PHStat allow the user to easily create stem-and-leaf displays for quantitative data. We demonstrate by using Example 2.2, found on pages 40-44 of the *Statistics for Business* and Economics text. A box plot for the 50 processing times is constructed below.

Example 2.5: Use PHStat to create a box plot for the 50 processing times of the data of Example 2.2.

Solution: We answer part b. by utilizing the Stem-and-Leaf Display utility within the **PHStat** program. Before we begin, we must access the data set for this example. **Open** the Data File **Example 2.2** which is found on the floppy disk included with this manual. Choose the **Box-and-Whisker Plot** option within the **PHStat** menu. Begin by entering the column where the data is located in the **Variable Cell Range** of this menu (see Figure 2.21). This can be done by **typing** the location or by **clicking and dragging** over the appropriate data cells in your workbook. You have the option of including the variable name in the first cell of data. If selected, the variable name will appear on the graph constructed. **Select** the **Single Group Variable** option whenever data appear in a single column. You also have the option of specifying a title for the box plot constructed. Click **OK**. The resulting box plot is shown in Figure 2.22.

Figure 2.21

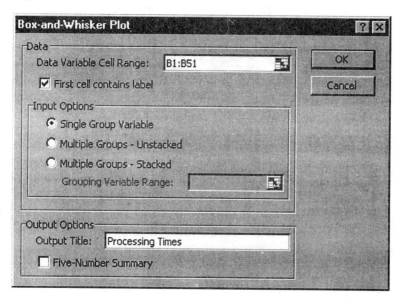

Figure 2.22

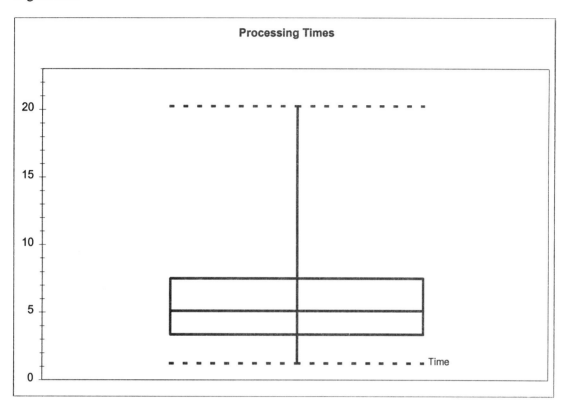

2.3 Numerical Techniques in Excel

2.3.1 Measures of Center

Excel allows the user to create many descriptive measure of data through the use of the Descriptive Statistics data analysis. While Excel doesn't distinguish between the different types of numerical measures, we choose to follow the *Statistics for Business and Economics* text and look at the measures of center, spread, and relative standing one at a time. We begin with measures of center.

Example 2.6   We use *Statistics for Business and Economics* Example 2.6 found on page 57 to illustrate the measures of center.

Calculate the sample mean for the R&D expenditure percentages of the 50 companies listed in Table 2.4.

Table 2.4

Company	Percentage	Company	Percentage	Company	Percentage
1	13.5	18	6.9	35	8.5
2	8.4	19	7.5	36	9.4
3	10.5	20	11.1	37	10.5
4	9	21	8.2	38	6.9
5	9.2	22	8	39	6.5
6	9.7	23	7.7	40	7.5
7	6.6	24	7.4	41	7.1
8	10.6	25	6.5	42	13.2
9	10.1	26	9.5	43	7.7
10	7.1	27	8.2	44	5.9
11	8	28	6.9	45	5.2
12	7.9	29	7.2	46	5.6
13	6.8	30	8.2	47	11.7
14	9.5	31	9.6	48	6
15	8.1	32	7.2	49	7.8
16	13.5	33	8.8	50	6.5
17	9.9	34	11.3		

Solution:

We must first retrieve data set to work with. If you are using Excel 97, **Open** the file SBE Table 2.3. Once the data is available, click on the **Tools** menu that appears at the top of the Excel worksheet. Select the **Data Analysis** option in the Tools menu and choose the **Descriptive Statistics** item (see Figure 2.23). Click **OK**.

Figure 2.23

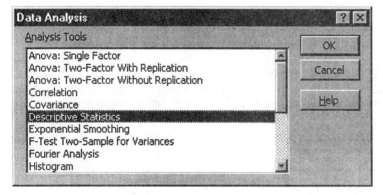

From the Descriptive Statistics menu, the user must specify the **Input Range**, the **Output Range**, and which statistics are desired. As with all Excel analyses, the Input Range should be the range of cells where the data set is located in the Excel worksheet. Either type or highlight with the mouse and enter the data set location for the Input Range (see Figure 2.24). The Output Range can either be a location within the current worksheet or a new Worksheet that you define. We opt to place the output in cell D1 of the current worksheet. Finally, the **Summary Statistics** box needs to be **checked** to generate the measures of center that are desired. Click **OK**.

Figure 2.24

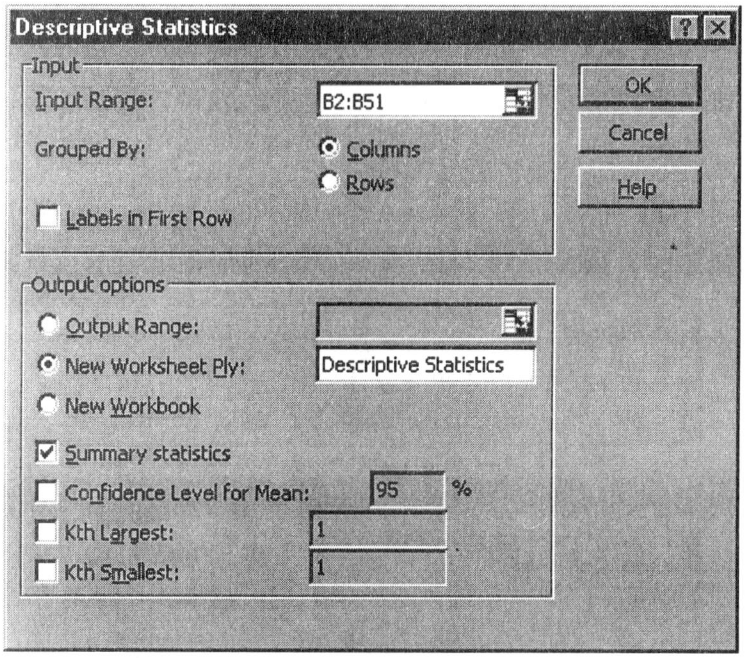

[Descriptive Statistics dialog box: Input Range B2:B51, Grouped By Columns, New Worksheet Ply: Descriptive Statistics, Summary statistics checked, Confidence Level for Mean 95%, Kth Largest 1, Kth Smallest 1]

Excel calculates the three measures of center, mean, median, and mode for the data set of interest (see Table 2.5). The mean R&D expenditure for the 50 companies is reported to be 8.492 percent.

Table 2.5

Column 1	
Mean	8.492
Standard Error	0.2801
Median	8.05
Mode	6.9
Standard Deviation	1.980604
Sample Variance	3.922792
Kurtosis	0.419288
Skewness	0.854601
Range	8.3
Minimum	5.2
Maximum	13.5
Sum	424.6
Count	50

Calculating the measures of center for other data sets requires only changing the Input Range values in the Descriptive Statistics menu above. Notice that the values for both the sample median and sample mode are also given on this printout.

2.3.2 Measures of Spread

The three measures of spread, range, standard deviation, and variance are found in the same manner as the measures of center above. We will use Example 2.9 from *Statistics for Business and Economics* to demonstrate.

Example 2.7 *Statistics for Business and Economics* Example 2.10 found of page 69.

Use the computer to find the sample variance s^2 and the sample standard deviation s for the 50 companies' percentage of revenues spent on R & D. The data is shown in Table 2.6.

Table 2.6

Company	Percentage	Company	Percentage	Company	Percentage
1	13.5	18	6.9	35	8.5
2	8.4	19	7.5	36	9.4
3	10.5	20	11.1	37	10.5
4	9	21	8.2	38	6.9
5	9.2	22	8	39	6.5
6	9.7	23	7.7	40	7.5
7	6.6	24	7.4	41	7.1
8	10.6	25	6.5	42	13.2
9	10.1	26	9.5	43	7.7
10	7.1	27	8.2	44	5.9
11	8	28	6.9	45	5.2
12	7.9	29	7.2	46	5.6
13	6.8	30	8.2	47	11.7
14	9.5	31	9.6	48	6
15	8.1	32	7.2	49	7.8
16	13.5	33	8.8	50	6.5
17	9.9	34	11.3		

Solution:

We must first retrieve a data set to work with. If you are using Excel 97, **Open** the file SBE Table 2.3. Once the data is available, click on the **Tools** menu that appears at the top of the Excel worksheet. Select the **Data Analysis** option in the Tools menu and choose the **Descriptive Statistics** item (see Figure 2.23). Click **OK**.

From the Descriptive Statistics menu, the user must specify the **Input Range**, the **Output Range**, and which statistics are desired. As with all Excel analyses, the Input Range should be the range of cells where the data set is located in the Excel worksheet. Either type or highlight with the mouse and enter the data set location for the Input Range (see Figure 2.24). The Output Range can either be a location within the current worksheet or a new Worksheet that you define. We opt to place the output in cell D1 of the current worksheet. Finally, the **Summary Statistics** box needs to be checked to generate the measures of center that are desired. Click **OK**.

Excel calculates the three measures of spread, range, standard deviation, and variance for the data set of interest (see Table 2.7). The sample variance for the R&D expenditure of the 50 companies is reported to be 3.922791837 and the sample standard deviation is 1.980603907 percent.

Table 2.7

Column 1	
Mean	8.492
Standard Error	0.2801
Median	8.05
Mode	6.9
Standard Deviation	1.980604
Sample Variance	3.922792
Kurtosis	0.419288
Skewness	0.854601
Range	8.3
Minimum	5.2
Maximum	13.5
Sum	424.6
Count	50

Calculating the measures of spread for other data sets requires only changing the Input Range values in the Descriptive Statistics menu above.

2.3.3 Measure of Relative Standing

Excel allows the user to calculate the two measures of relative standing, percentiles and z-scores through the use of two of it's many functions. We first look at how Excel calculates percentiles.

Example 2.8 *Statistics for Business and Economics* Example 2.9 found of page 69.

Use the data from Example 2.9 to calculate the 20th percentile of the R&D percentages.

Solution:

We must first retrieve a data set to work with. If you are using Excel 97, **Open** the file SBE Table 2.3. Once the data is available, click the f_x **icon** at the top of the Excel worksheet. Choose the **Statistical Function Category** and cursor down until you reach the function name **PERCENTILE** (see Figure 2.25). The PERCENTILE function has the form:

PERCENTILE(array,k)

where **array** represents the location of the data set that you want to find the percentile for, and **k** is a number between 0 and 1 that represents the percentile that is desired.

56 Chapter 2: Methods for Describing Sets of Data

Figure 2.25

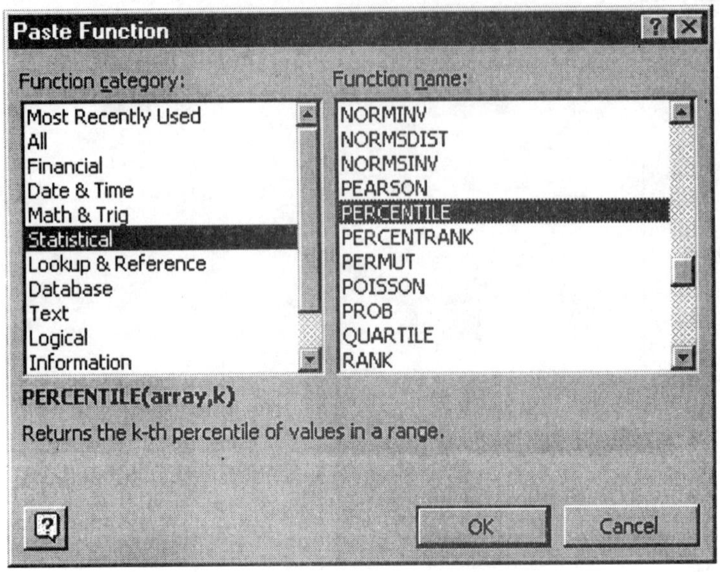

![Paste Function dialog box with Statistical category selected and PERCENTILE function highlighted. PERCENTILE(array,k) - Returns the k-th percentile of values in a range.]

For this example, the 50 R&D percentages are located in column B in rows 2 through 51. We assign the **Array** location to be B2:B51 (see Figure 2.26). We also assign the value of **K** to be .20 respresenting the 20th percentile. Click **OK**.

Figure 2.26

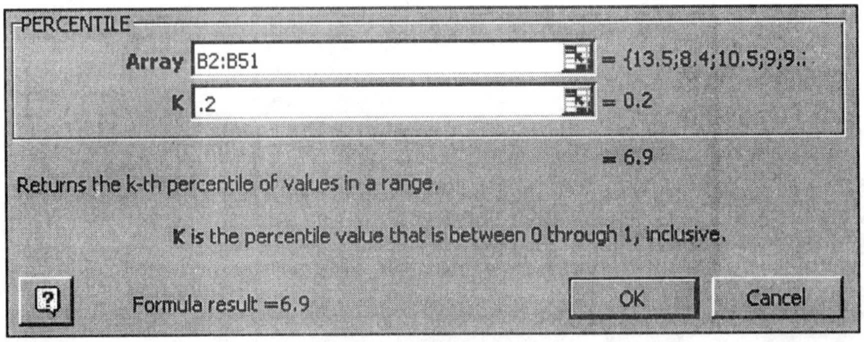

![PERCENTILE function dialog box. Array: B2:B51 = {13.5;8.4;10.5;9;9.:}, K: .2 = 0.2, = 6.9. Returns the k-th percentile of values in a range. K is the percentile value that is between 0 through 1, inclusive. Formula result =6.9]

Excel returns a value of 6.9. We interpret that 6.9 represents the 20th percentile of the 50 R&D percentages in our data set. By changing the data set and the value of K, we can find percentiles for any group of data.

The second measure of relative standing is the z-score. Again, we turn in Excel to a function that will allow the user to calculate values for a z-score. For purposes of illustration, we will again use the data from Example 2.9 to find a z-value.

Example 2.9 Statistics for Business and Economics Example 2.9 found on page 69.

Use the 50 R&D percentages to find the z-score for an R&D percentage of 10%.

Solution:

We must first retrieve a data set to work with. If you are using Excel 97, **Open** the file SBE Table 2.3. Once the data is available, click the f_x **icon** at the top of the Excel worksheet. Choose the

Section 2.3: Numerical Techniques In Excel

Statistical Function Category and cursor down until you reach the function name **STANDARDIZE** (see Figure 2.27). The STANDARDIZE function has the form:

STANDARDIZE (x, mean, standardize_dev)

where   **x** represents the value that you wish to determine the z-score for,
        mean represents the mean of the data set that you want to find the z-score for, and
        standard deviation represents the standard deviation of the data set that you want to find the z-score for.

Figure 2.27

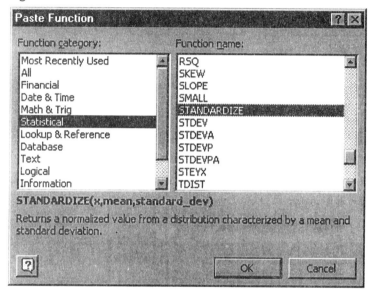

For this example, we use the value of 10 as our choice for **X** in the STANDARDIZE function (see Figure 2.28). From the work we did on Example 2.9, we know to use a value of 8.492 for the **mean** and a value of 1.980603907 for the **standard deviation**. Click **OK**.

Figure 2.28

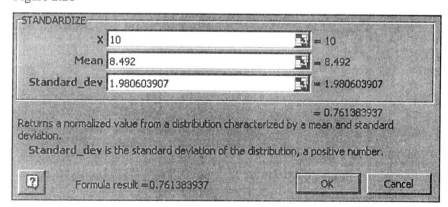

Excel returns a value of 0.762383937. We make the interpretation that an R&D percentage of 10% would fall approximately .76 standard deviation above the mean R&D percentage of the 50 companies. By changing the values of X, Mean, and Standard Deviation, we can find z-scores for a wide variety of situations.

Chapter 2: Methods for Describing Sets of Data

Technology Lab

Utilize the data from Exercise 2.107 in the text to practice the Excel skills that were taught in this chapter.

A manufacturer of industrial wheels is losing many profitable orders because of the long time it takes the firm's marketing, engineering, and accounting departments to develop price quotes for potential customers. To remedy this problem the firm's management would like to set guidelines for the length of time each department should spend developing price quotes. To help develop these guidelines, 50 requests for price quotes were randomly selected from the set of all price quotes made last year; the processing time was determined for each department. These times are displayed in the table below and are contained in the Excel file Exercise 2_107. The price quotes are also classified by whether they were "lost" (i.e., whether or not the customer placed an order after receiving the price quote).

a. Construct a bar graph and a pie chart to determine the relative frequencies of the orders that were lost.
b. Construct a histogram for the price quote processing times (in days) of the marketing department.
c. Construct a stem-and-leaf display for the price quote processing times (in days) of the engineering department.
d. Construct a scatterplot that compares the price quote processing times (in days) of the engineering and the accounting departments.
e. Construct a box plot for the price quote processing times (in days) of the marketing department.
f. Find the descriptive statistics for the price quote processing times (in days) of the all three departments.

Use the Excel output provided below to check your work.

PRICE QUOTE PROCESSING TIME (IN Days)

Request Number	Marketing	Engineering	Accounting	Lost?	Request Number	Marketing	Engineering	Accounting	Lost?
1	7	6.2	0.1	2	26	0.6	2.2	0.5	2
2	0.4	5.2	0.1	2	27	6	1.8	0.2	2
3	2.4	4.6	0.6	2	28	5.8	0.6	0.5	2
4	6.2	13	0.8	1	29	7.8	7.2	2.2	1
5	4.7	0.9	0.5	2	30	3.2	6.9	0.1	2
6	1.3	0.4	0.1	2	31	11	1.7	3.3	2
7	7.3	6.1	0.1	2	32	6.2	1.3	2	2
8	5.6	3.6	3.8	2	33	6.9	6	10.5	1
9	5.5	9.6	0.5	2	34	5.4	0.4	8.4	2
10	5.3	4.8	0.8	2	35	6	7.9	0.4	2
11	6	2.6	0.1	2	36	4	1.8	18.2	1
12	2.6	11.3	1	2	37	4.5	1.3	0.3	2
13	2	0.6	0.8	2	38	2.2	4.8	0.4	2
14	0.4	12.2	1	2	39	3.5	7.2	7	1
15	8.7	2.2	3.7	2	40	0.1	0.9	14.4	2
16	4.7	9.6	0.1	2	41	2.9	7.7	5.8	2
17	6.9	12.3	0.2	1	42	5.4	3.8	0.3	2
18	0.2	4.2	0.3	2	43	6.7	1.3	0.1	2
19	5.5	3.5	0.4	2	44	2	6.3	9.9	1
20	2.9	5.3	22	2	45	0.1	12	3.2	2
21	5.9	7.3	1.7	2	46	6.4	1.3	6.2	2
22	6.2	4.4	0.1	2	47	4	2.4	13.5	1
23	4.1	2.1	30	1	48	10	5.3	0.1	2
24	5.8	0.6	0.1	2	49	8	14.4	1.9	1
25	5	3.1	2.3	2	50	7	10	2	2

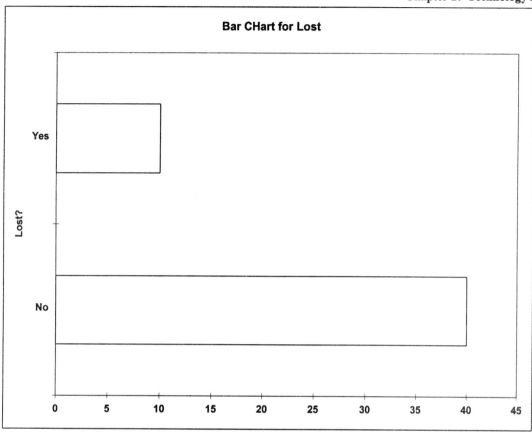

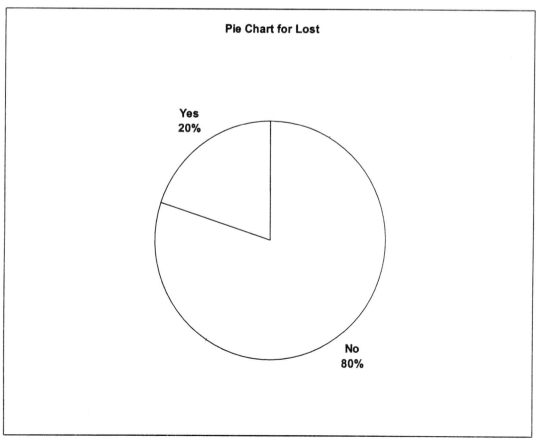

60 Chapter 2: Methods for Describing Sets of Data

```
Engineering Processing Times
for Engineering
Stem unit:    1

         0|4 4 6 6 6 9 9
         1|3 3 3 3 7 8 8
         2|1 2 2 4 6
         3|1 5 6 8
         4|2 4 6 8 8
         5|2 3 3
         6|0 1 2 3 9
         7|2 2 3 7 9
         8|
         9|6 6
        10|0
        11|3
        12|0 2 3
        13|0
        14|4
```

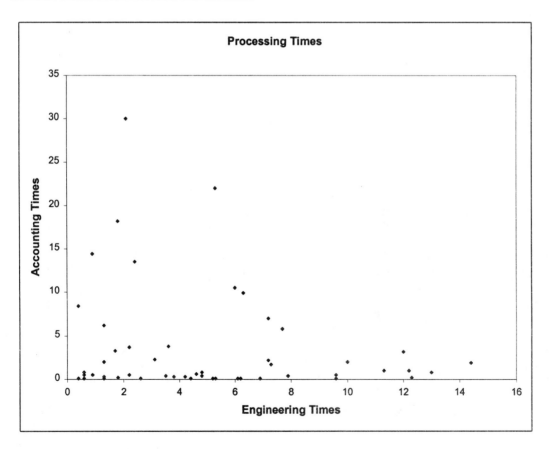

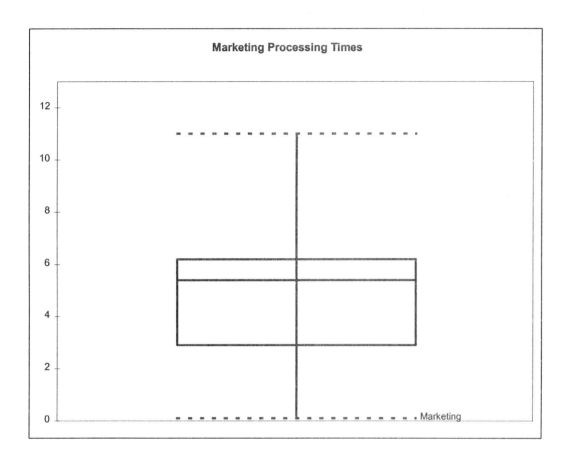

	Marketing	Engineering	Accounting
Mean	4.766	5.044	3.652
Standard Error	0.3654308	0.54229414	0.88479325
Median	5.4	4.5	0.8
Mode	6.2	1.3	0.1
Standard Deviation	2.583986	3.83459867	6.2564331
Sample Variance	6.6769837	14.7041469	39.1429551
Kurtosis	-0.2561865	-0.3393416	6.90660771
Skewness	-0.1027209	0.75624587	2.55214148
Range	10.9	14	29.9
Minimum	0.1	0.4	0.1
Maximum	11	14.4	30
Sum	238.3	252.2	182.6
Count	50	50	50

Chapter 3
Probability

3.1 Introduction

Chapter 3 introduces the topic of probability and random sampling to the reader. PHStat allows the user to work with simple probabilities found in a 2x2 table. In addition, PHStat offers the reader a convenient method of drawing a sample of size n from a population of N items. Both procedures are discussed below.

3.2 Probabilities in a 2x2 Table

Chapter 3 introduces the reader to the idea of presenting descriptive results of collected data in a tabular form. The simplest form of these tables is the 2x2 table, where data from two variables (with two levels each) are presented. An example of a 2x2 table is shown below in Table 3.1. Table 3.1 shows data collected from a sample of customers at a neighborhood restaurant. The two variables shown in the table are age of customer and sex of customer

Table 3.1

		Age of Customer	
		18-35	36-70
Sex	**Male**	40	12
	Female	31	17

PHStat offers the reader a probability utility that will allow the calculation of simple probabilities, as well as the probabilities of unions and intersections. This utility is only offered for data that can be arranged in a 2x2 table similar to the one shown in Table 3.1. We illustrate with the following example.

Example 3.1 Use the data shown in Table 3.1 to find the following:
   a. Find the probability that a randomly selected customer is male.
   b. Find the proportion of all customers that are female or aged 36-70.
   c. Find the probability that a randomly selected customer is male and 18-35.

To use the probability function within PHStat, **open** a new workbook and place the cursor in the upper left cell of the worksheet. **Click** on the **PHStat** menu at the top of the screen. **Select** the **Data Preparation** option from the choices available and then select the **Probabilities** option from those listed. You should open a worksheet that looks like the one shown in Figure 3.1.

This worksheet is a template that allows the user to change the values of the sample space to represent the data of their 2x2 table. We begin by replacing the generic labels **Event A** and **Event B** with the variable names of our example, **Age** and **Sex of Customer**. The next step is to replace the outcomes that are listed in the table as A1, A2, B1, and B2 with the outcomes that are meaningful in our example (e.g., male, female, 18-35, and 36-70). The final step is to change the numbers shown in the table with the numbers that are shown in Table 3.1. The changed worksheet is shown in Figure 3.2

64   Chapter 3: Probability

Figure 3.1

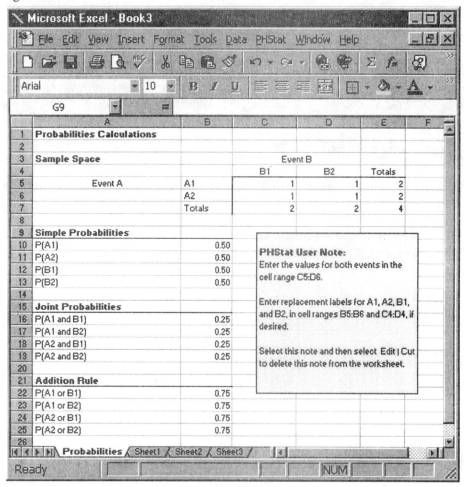

Figure 3.2

Probabilities Calculations					
Sample Space			Age of Customer		
			18-35	36-70	Totals
Sex		Male	40	12	52
		Female	31	17	48
		Totals	71	29	100
Simple Probabilities					
P(Male)	0.52				
P(Female)	0.48				
P(18-35)	0.71				
P(36-70)	0.29				
Joint Probabilities					
P(Male and 18-35)	0.40				
P(Male and 36-70)	0.12				
P(Female and 18-35)	0.31				
P(Female and 36-70)	0.17				
Addition Rule					
P(Male or 18-35)	0.83				
P(Male or 36-70)	0.69				
P(Female or 18-35)	0.88				
P(Female or 36-70)	0.60				

The answers to the questions in Example 3.1 can each be found in the probabilities calculated in the worksheet. The probability that a randomly selected customer is male is shown to be .52. The proportion of all customers that are female or aged 36-70 is shown to be .60. The probability that a randomly selected customer is male and 18-35 is .40. Other similar probabilities are shown in the table as well. [Note to reader: The data for this example can be found in the data set Table 3-1 found on the accompanying data disk]

3.3 Random Sampling

Section 3.7 in the text defines to the reader what a random sample is and gives a method of generating a random sample that utilizes the random number table found in Table 1 of Appendix B. PHStat offers a simpler method of drawing a random sample from a population of known population size.

To use the random sample generator within PHStat, **open** a new workbook and place the cursor in the upper left cell of the worksheet. **Click** on the **PHStat** menu at the top of the screen. **Select** the **Data Preparation** option from the choices available and then select the **Random Sample Generator** option from those listed. You will open the Random Sample Generator Menu shown in Figure 3.3. PHStat offers two methods of drawing a random sample. The first involves the user selecting both the sample size and population size that is appropriate for a particular sampling problem. The second involves actually selecting the sample size from a set of data that is already entered into an Excel worksheet. We illustrate both methods on the following page.

66 Chapter 3: Probability

Figure 3.3

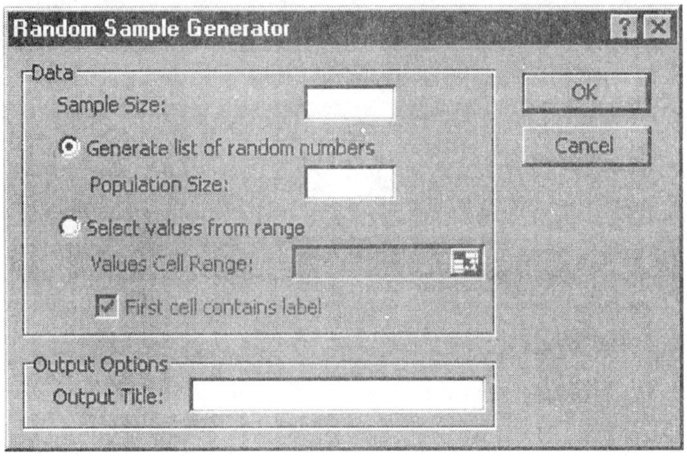

Example 3.2: Find which objects to sample if you wish to randomly sample 10 items from a list of 400 items.

To solve this problem, we simply enter the value **10** for the sample size and **enter** the value **400** for the population size in the menu above (see Figure 3.4). We have the choice of **entering a title** for the resulting output if we desire. When finished we **click** on **OK**. The output is shown below in Table 3.2. It is important to note that if we again use the same values for n and N, the sample items will change.

Figure 3.4

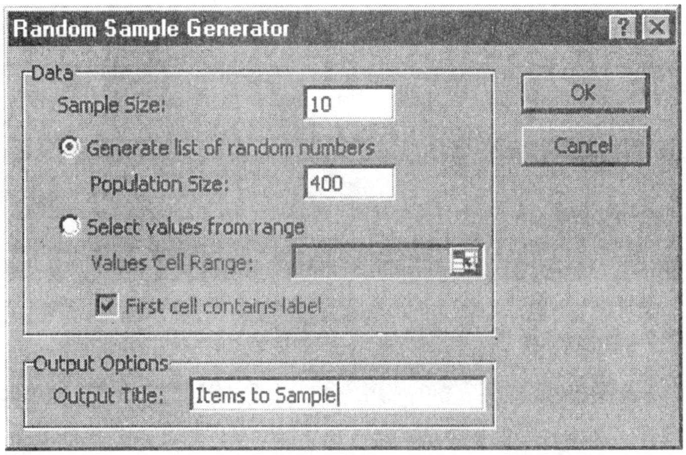

Section 3.3: Random Sampling

Table 3.2

Items to Sample
341
113
309
156
370
29
221
319
236
212

Example 3.3: Use the data set from Example 2.18 to randomly select five Length of Stays from the 50 that were given in the data set.

We first open data set SBE Example 2_18. We again must **enter** the sample size that is desired; in this case the value **5** will be entered. For this use of the random sample generator, we **select the select values from range** option. We then **enter the cell location** of the data that we wish to sample from, in this case the Length of Stay data located in cells **B2 through B52**. Again, a **title** can be entered if desired (see Figure 3.5). **Click OK**. The output is shown below in Table 3.3.

Figure 3.5

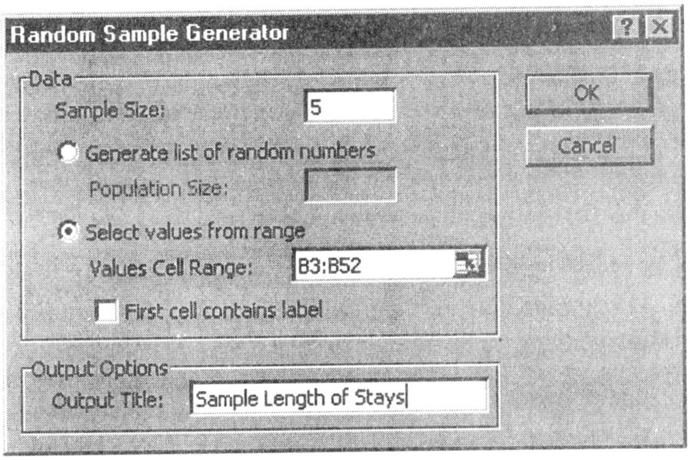

[Random Sample Generator dialog box: Sample Size: 5; Select values from range; Values Cell Range: B3:B52; Output Title: Sample Length of Stays]

Table 3.3

Sample Length of Stays
7
6
8
12
4

Note again that if we were to use the random sample generator again, the resulting output would differ, as the results of the random numbers selected would vary (randomly) from one sample to the next.

Chapter 4
Discrete Random Variables

4.1 Introduction

Chapter 4 introduces the two types of random variables, discrete and continuous, and discusses the discrete random variables in detail. Two discrete random variables, the binomial and poisson, are introduced and discussed in sections 4.4 and 4.5, respectively. The text introduces two main methods of working with both the binomial and poisson random variables. The first is to use their corresponding probability formulas (given in the text) to calculate probabilities associated with the random variables. The second is to use the probability tables reproduced in the appendix of the text. These tables are designed to give cumulative probabilities for specific binomial and poisson distributions.

Through the use of it's probability distributions, PHStat can be used to find both individual and cumulative probabilities for both the binomial and poisson random variables. PHStat can be used in place of either the formula or table methods mentioned above. The Cumulative Probability option that PHStat offers allows the user to find a variety of forms for the probabilities of interest. Whether the problems asks the reader for "at least", "at most", "more than", "less than", or "equal to", PHStat will provide the desired probability for both binomial and poisson probability distributions.

The following examples from *Statistics for Business and Economics* are solved with Microsoft Excel® in this chapter:

Excel Companion		Statistic for Business and Economics	
Example	**Page**	**SBE**	**SBE Page**
4.1	67	Example 4.11	194
4.2	69	Example 4.12	202

4.2 Calculating Binomial Probabilities

To use the binomial probability tool within PHStat, **open** a new workbook and place the cursor in the upper left cell of the worksheet. **Click** on the **PHStat** menu at the top of the screen. **Select** the **Probability Distribution** option from the choices available and then select the **Binomial** option from those listed. You should open the Binomial Probability Distribution menu that looks like the one shown in Figure 4.1.

Chapter 4: Discrete Random Variables

Figure 4.1

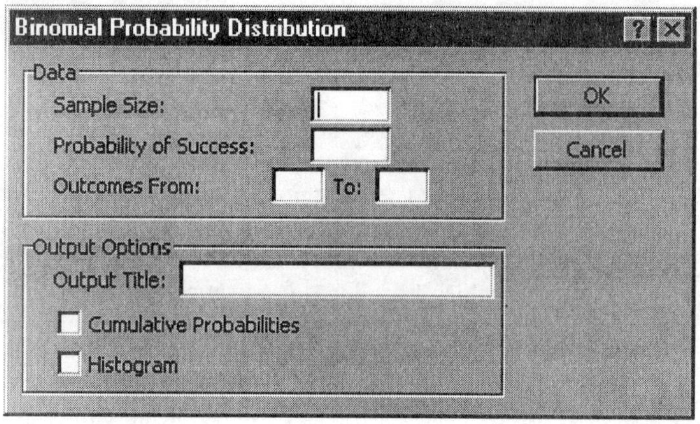

```
Binomial Probability Distribution                    [?][X]
┌─Data─────────────────────────────┐  ┌────────┐
│ Sample Size:        [        ]   │  │   OK   │
│ Probability of Success: [    ]   │  └────────┘
│ Outcomes From: [  ] To: [  ]     │  ┌────────┐
└──────────────────────────────────┘  │ Cancel │
┌─Output Options───────────────────┐  └────────┘
│ Output Title: [              ]   │
│ ☐ Cumulative Probabilities       │
│ ☐ Histogram                      │
└──────────────────────────────────┘
```

The user is required to enter the **sample size**, n, and the **probability of success**, p, from the binomial probability distribution of interest. The user then must specify the values of the **Outcomes** that he or she wishes to find probabilities for. For most applications, the **Cumulative Probabilities** option should be selected in order to maximize the information that PHStat will offer. An **Output Title** can be optionally selected and a **Histogram** can be specified if the user so desires. **Click OK** to finish. We illustrate with the next example.

Example 4.1: As an example, we turn to Example 4.11 from *Statistics for Business and Economics* found on page 194 of the text.

Suppose a poll of 20 employees is taken in a large company. The purpose is to determine x, the number who favor unionization. Suppose that 60% of all the company's employees favor unionization.

  a. Find the mean and standard deviation of x.
  b. Find the probability that x < 10.
  c. Find the probability that x > 12.
  d. Find the probability that x = 11.

Solution:

We utilize the **Binomial Probability Distribution** within PHStat to solve parts b-d. In order to solve these questions, we identify in the problem that the sample size is **n=20** and the probability of a success is **p=.60**. We enter both of these values in the appropriate locations in the **Binomial Probability Distribution** menu shown if Figure 4.2. We choose here to specify all possible **Outcomes 0 to 20** in this problem to demonstrate the output that PHStat is capable of. Note that to answer parts b-d, we only need to specify the Outcomes 10 to 12. We also **check** the **Cumulative Probabilities** box and finish by **clicking OK**. The output generated by PHStat is shown in Table 4.1.

Section 4.2: Calculating Binomial Probabilities

Figure 4.2

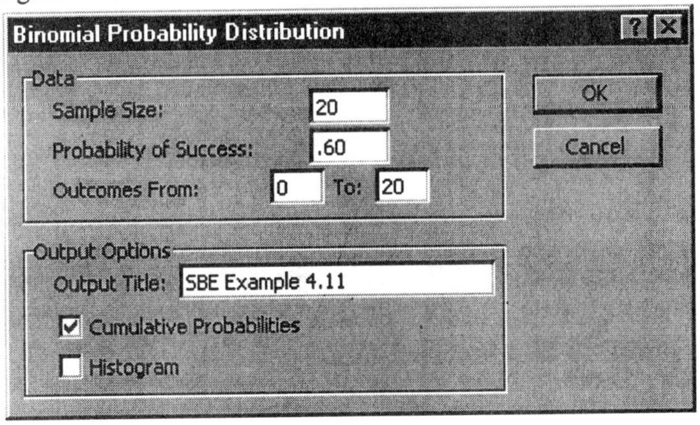

Binomial Probability Distribution

Data
- Sample Size: 20
- Probability of Success: .60
- Outcomes From: 0 To: 20

Output Options
- Output Title: SBE Example 4.11
- ☑ Cumulative Probabilities
- ☐ Histogram

[OK] [Cancel]

Table 4.1

SBE Example 4.11

Sample size	20
Probability of success	0.6
Mean	12
Variance	4.8
Standard deviation	2.19089

Binomial Probabilities Table

X	P(X)	P(<=X)	P(<X)	P(>X)	P(>=X)
0	1.1E-08	1.1E-08	0	1	1
1	3.3E-07	3.41E-07	1.1E-08	1	1
2	4.7E-06	5.04E-06	3.41E-07	0.999995	1
3	4.23E-05	4.73E-05	5.04E-06	0.999953	0.999995
4	0.00027	0.000317	4.73E-05	0.999683	0.999953
5	0.001294	0.001612	0.000317	0.998388	0.999683
6	0.004854	0.006466	0.001612	0.993534	0.998388
7	0.014563	0.021029	0.006466	0.978971	0.993534
8	0.035497	0.056526	0.021029	0.943474	0.978971
9	0.070995	0.127521	0.056526	0.872479	0.943474
10	0.117142	0.244663	**0.127521**	0.755337	0.872479
11	**0.159738**	0.404401	0.244663	0.595599	0.755337
12	0.179706	0.584107	0.404401	**0.415893**	0.595599
13	0.165882	0.749989	0.584107	0.250011	0.415893
14	0.124412	0.874401	0.749989	0.125599	0.250011
15	0.074647	0.949048	0.874401	0.050952	0.125599
16	0.034991	0.984039	0.949048	0.015961	0.050952
17	0.01235	0.996389	0.984039	0.003611	0.015961
18	0.003087	0.999476	0.996389	0.000524	0.003611
19	0.000487	0.999963	0.999476	3.66E-05	0.000524
20	3.66E-05	1	0.999963	0	3.66E-05

72 Chapter 4: Discrete Random Variables

The answers to questions b-d are shown in the highlighted boxes above. Notice the column that each appears in. PHStat can be used to solve a wide variety of binomial probabilities by simply changing the values of n, p, and the outcomes desired.

4.3 Calculating Poisson Probabilities

To use the poisson probability tool within PHStat, **open** a new workbook and place the cursor in the upper left cell of the worksheet. **Click** on the **PHStat** menu at the top of the screen. **Select** the **Probability Distribution** option from the choices available and then select the **Poisson** option from those listed. You should open the Poisson Probability Distribution menu that looks like the one shown in Figure 4.3.

Figure 4.3

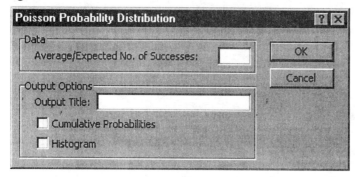

The user is required to enter the **average or expected number of successes** (e.g., the mean), λ, from the poisson probability distribution of interest. For most applications, the **Cumulative Probabilities** option should be selected in order to maximize the information that PHStat will offer. An **Output Title** can be optionally selected and a **Histogram** can be specified if the user so desires. **Click OK** to finish. We illustrate with the next example.

Example 4.2: As an example, we turn to Example 4.12 from *Statistics for Business and Economics* found on page 202 of the text:

Suppose the number, x, of a company's employees who are absent on Mondays has (approximately) a Poisson probability distribution. Furthermore, assume that the average number of Monday absentees is 2.6.

a. Find the mean and standard deviation of x, the number of employees absent on Monday.
b. Find the probability that fewer than two employees are absent on a given Monday.
c. Find the probability that more than five employees are absent on a given Monday.
d. Find the probability that exactly five employees are absent on a given Monday.

Solution:

We utilize the **Poisson Probability Distribution** within PHStat to solve parts b-d. In order to solve these questions, we identify in the problem that the mean absenteeism within the problem is $\lambda=2.6$. We enter this value in the appropriate location in the **Poisson Probability Distribution** menu shown if Figure 4.4 We **check** the **Cumulative Probabilities** box and finish by **clicking OK**. The output generated by PHStat is shown in Table 4.2.

Figure 4.4

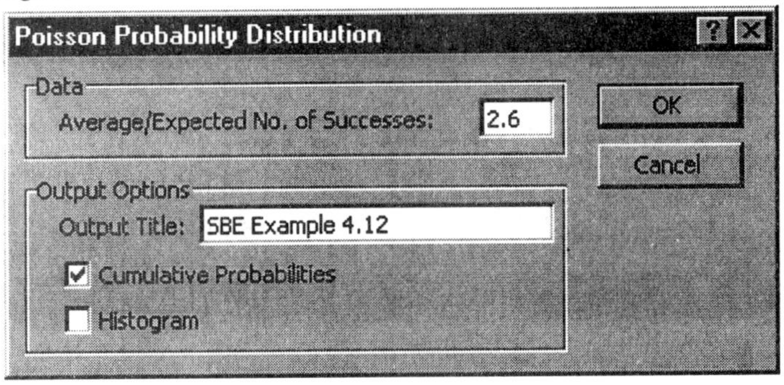

Table 4.2

SBE Example 4.12

Average/Expected number of successes: 2.6

Poisson Probabilities Table

X	P(X)	P(<=X)	P(<X)	P(>X)	P(>=X)
0	0.074274	0.074274	0.000000	0.925726	1.000000
1	0.193111	0.267385	0.074274	0.732615	0.925726
2	0.251045	0.518430	0.267385	0.481570	0.732615
3	0.217572	0.736002	0.518430	0.263998	0.481570
4	0.141422	0.877423	0.736002	0.122577	0.263998
5	0.073539	0.950963	0.877423	0.049037	0.122577
6	0.031867	0.982830	0.950963	0.017170	0.049037
7	0.011836	0.994666	0.982830	0.005334	0.017170
8	0.003847	0.998513	0.994666	0.001487	0.005334
9	0.001111	0.999624	0.998513	0.000376	0.001487
10	0.000289	0.999913	0.999624	0.000087	0.000376
11	0.000068	0.999982	0.999913	0.000018	0.000087
12	0.000015	0.999996	0.999982	0.000004	0.000018
13	0.000003	0.999999	0.999996	0.000001	0.000004
14	0.000001	1.000000	0.999999	0.000000	0.000001
15	0.000000	1.000000	1.000000	0.000000	0.000000
16	0.000000	1.000000	1.000000	0.000000	0.000000
17	0.000000	1.000000	1.000000	0.000000	0.000000
18	0.000000	1.000000	1.000000	0.000000	0.000000
19	0.000000	1.000000	1.000000	0.000000	0.000000
20	0.000000	1.000000	1.000000	0.000000	0.000000

The answers to questions b-d are shown in the highlighted boxes above. Notice the column that each appears in. PHStat can be used to solve a wide variety of poisson probabilities by simply changing the value of λ.

Technology Lab

The following two exercises have been taken from Statistics for Business and Economics for you to practice the techniques discussed in this chapter. The output generated from these problems is also given for you to check your work.

4.83 The efficacy of insecticides is often measured by the dose necessary to kill a certain percentage of insects. Suppose a certain dose of a new insecticide is supposed to kill 80% of the insects that receive it. To test the claim, 25 insects are exposed to the insecticide.

   a. If the insecticide really kills 80% of the exposed insects, what is the probability that fewer than 15 die?

PHStat Output

Binomial Probabilities

Sample size	25
Probability of success	0.8
Mean	20
Variance	4
Standard deviation	2

Binomial Probabilities Table

X	P(X)	P(<=X)	P(<X)	P(>X)	P(>=X)
14	0.004015	0.005555	0.00154	0.994445	0.99846
15	0.011777	0.017332	**0.005555**	0.982668	0.994445

4.84 Large bakeries typically have fleets of delivery trucks. One such bakery determined that the expected number of delivery truck breakdowns per day is 1.5. Assume that the number of breakdowns is independent from day to day.

   a. What is the probability that there will be exactly two breakdowns today and exactly three breakdowns tomorrow?
   b. Fewer than two today and more than two tomorrow?

PHStat Output

Poisson Probabilities for Customer Arrivals

Average/Expected number of successes:	1.5

Poisson Probabilities Table

X	P(X)	P(<=X)	P(<X)	P(>X)	P(>=X)
0	0.223130	0.223130	0.000000	0.776870	1.000000
1	0.334695	0.557825	0.223130	0.442175	0.776870
2	**0.251021**	0.808847	**0.557825**	**0.191153**	0.442175
3	**0.125511**	0.934358	0.808847	0.065642	0.191153
4	0.047067	0.981424	0.934358	0.018576	0.065642

Chapter 5
Continuous Random Variables

5.1 Introduction

Chapter 5 introduces the second types of random variables, continuous random variables, and discusses the uniform, normal and exponential random variables in detail. The text introduces statistical tables as the best methods of working with the normal and the exponential random variables. The uniform random variable can be solved rather easily using simple mathematics.

PHStat can be used in place of the statistical tables to find probabilities for both the normal and exponential distributions. In addition, PHStat offers the user the ability to assess the normality of a distribution of data through the use of its normal probability plot option. The following examples from *Statistics for Business and Economics* are solved with PHStat in this chapter.

Excel Companion		Statistics for Business and Economics	
Example	Page	SBE	Page
5.1	76	Example 5.7	227
5.2	77	Example 5.10	229
5.3	78	Table 5.2	235
5.4	80	Example 5.13	248

5.2 Calculating Normal Probabilities

To use the normal probability tool within PHStat, **open** a new workbook and place the cursor in the upper left cell of the worksheet. **Click** on the **PHStat** menu at the top of the screen. **Select** the **Probability Distribution** option from the choices available and then select the **Normal** option from those listed. You should open the Normal Probability Distribution menu that looks like the one shown in Figure 5.1.

Figure 5.1

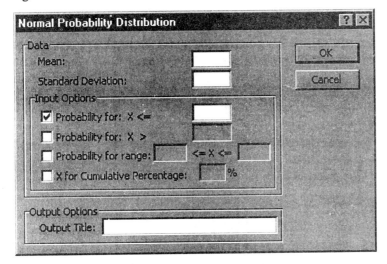

76 Chapter 5: Continuous Random Variables

The user is required to enter the **Mean**, μ, and the **Standard Deviation**, σ, from the normal probability distribution of interest. The user then must specify the type of Input Options that they desire. Several different probability options are available as is finding a specified value of X in the normal distribution. An **Output Title** can be optionally selected if the user so desires. **Click OK** to finish. We illustrate with the next two examples.

Example 5.1: We use Example 5.7 from *Statistics for Business and Economics* found on page 227 in the text:

Suppose an automobile manufacturer introduces a new model that has an advertised mean in-city mileage of 27 miles per gallon. Although such advertisements seldom report in any measure of variability, suppose you write the manufacturer for the details of the test, and find that the standard deviation is 3 miles per gallon. This information leads you to formulate a probability model for the random variable, x, the in-city mileage for this car model. You believe that the probability distribution of x can be approximated by a normal distribution with a mean of 27 and a standard deviation of 3.

a. If you were to buy this model of automobile, what is the probability that you would purchase one that averages less than 20 miles per gallon for in-city driving? In other words, find $P(x < 20)$.

Solution:

We utilize the **Normal Probability Distribution** within PHStat to solve the problem. In order to solve these questions, we identify in the problem that the mean is $\mu=27$ and the standard deviation is $\sigma=3$. We **enter** both of these values in the appropriate locations in the **Normal Probability Distribution** menu shown if Figure 5.2. We choose here to **specify** the **Input Option Probability for: X <=** as the question we wish to solve is $P(x < 20)$. We **enter** the value **20** in the appropriate space in the menu. We also **enter** an **Output Title** and finish by **clicking OK**. The output generated by PHStat is shown in Table 5.1.

Figure 5.2

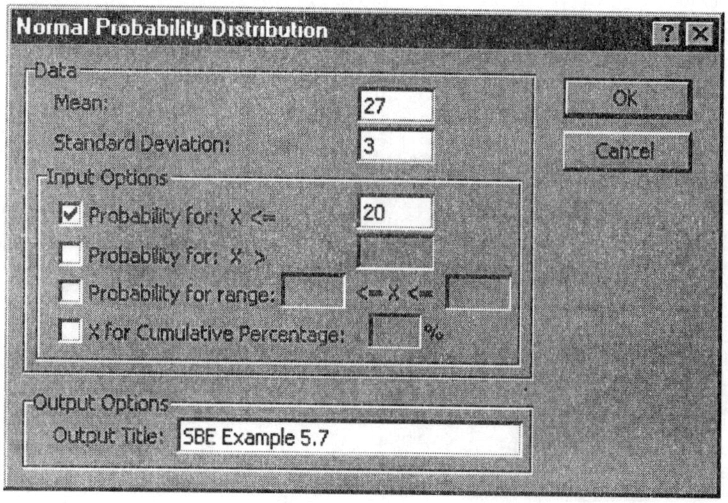

Section 5.2: Calculating Normal Probabilities 77

Table 5.1

SBE Example 5.7	
Mean	27
Standard Deviation	3
Probability for X<=	20
Z Value	-2.333333333
P(X<=20)	0.009815307

PHStat returns the probability 0.0098. We can verify this result using the normal procedures discussed in the text and Table IV of Appendix B. By changing the input options in the normal probability distribution menu, the user can calculate any type of probability desired. We illustrate finding a point in the normal distribution in the next example.

Example 5.2: We use the Example 5.10 from *Statistics for Business and Economics* found on page 229 in the text:

Suppose a paint manufacturer has a daily production, x, that is normally distributed with a mean of 100,000 gallons and a standard deviation of 10,000 gallons. Management wants to create an incentive bonus for the production crew when the daily production exceeds the 90th percentile of the distribution, in hopes that the crew will, in turn, become more productive. At what level of production should management pay the incentive bonus?

Solution:

We utilize the **Normal Probability Distribution** within PHStat to solve the problem. In order to solve these questions, we identify in the problem that the mean is µ=100,000 and the standard deviation is σ=10,000. We **enter** both of these values in the appropriate locations in the **Normal Probability Distribution** menu shown if Figure 5.3. We choose here to **specify** the **Input Option X for Cumulative Percentage:** as the question we wish to find the 90th percentile of the production distribution. We **enter** the value **90** in the appropriate space in the menu. We also **enter** an **Output Title** and finish by **clicking OK**. The output generated by PHStat is shown in Table 5.2.

Figure 5.3

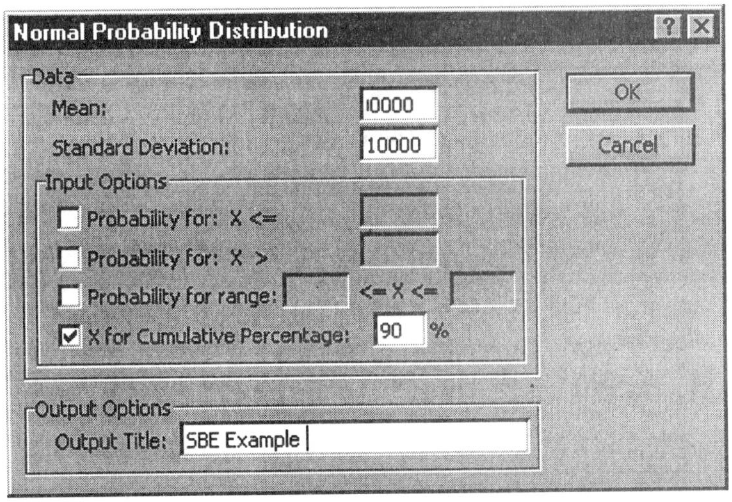

78 Chapter 5: Continuous Random Variables

Table 5.2

SBE Example	
Mean	100000
Standard Deviation	10000
Find X and Z	
Cumulative Percentage:	90.00%
Z Value	1.281550794
X Value	112815.5079

PHStat returns the desired production level of 112,815 gallons of paint. We can verify this result using the normal procedures discussed in the text and Table IV of Appendix B. By changing the cumulative percentage in the normal probability distribution menu, the user can find any value of X desired.

5.3 Assessing the Normality of a Data Set

PHStat offers the user a method of assessing whether a data set possesses a normal distribution. The Normal Probability Plot utility creates a plot that enables the reader to determine the shape of the data. The Normal Probability Plot is found by **clicking** on the **PHStat** menu and **selecting** the **Probability Distribution** option. Select the **Normal Probability Plot** option to generate a menu similar to the one shown below in Figure 5.4. The user must **specify** the **Variable Cell Range** that the data is located in and has the option of titling the output. **Click OK** when finished. We illustrate below with the following example.

Figure 5.4

Example 5.3: We utilize the EPA Gas Mileage Ratings for 100 Cars that is given in Table 5.2 of the *Statistics for Business and Economics* text found on page 235. The data is shown below in Table 5.3. Construct a normal probability plot of the data and assess the shape of the EPA gas mileage ratings.

Section 5.3: Assessing the Normality of a Data Set

Table 5.3

EPA Gas Mileage Ratings for 100 Cars (miles per gallon)									
36.3	41.0	36.9	37.1	44.9	36.8	30.0	37.2	42.1	36.7
32.7	37.3	41.2	36.6	32.9	36.5	33.2	37.4	37.5	33.6
40.5	36.5	37.6	33.9	40.2	36.4	37.7	37.7	40.0	34.2
36.2	37.9	36.0	37.9	35.9	38.2	38.3	35.7	35.6	35.1
38.5	39.0	35.5	34.8	38.6	39.4	35.3	34.4	38.8	39.7
36.3	36.8	32.5	36.4	40.5	36.6	36.1	38.2	38.4	39.3
41.0	31.8	37.3	33.1	37.0	37.6	37.0	38.7	39.0	35.8
37.0	37.2	40.7	37.4	37.1	37.8	35.9	35.6	36.7	34.5
37.1	40.3	36.7	37.0	33.9	40.1	38.0	35.2	34.8	39.5
39.9	36.9	32.9	33.8	39.8	34.0	36.8	35.0	38.1	36.9

Solution:

We first open the Excel data set SBE Table 5-2 to access the worksheet that contains the EPA gas mileage ratings. We **specify** the location of the data in the **Variable Cell Range** portion of the Normal Probability Plot menu (we choose to contain the label in the first cell of the specified range). We also choose to select an **Output Title** for the plot. To finish, we **click OK** (see Figure 5.5). The plot created by PHStat is shown in Figure 5.6.

Figure 5.5

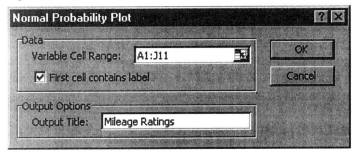

Figure 5.6

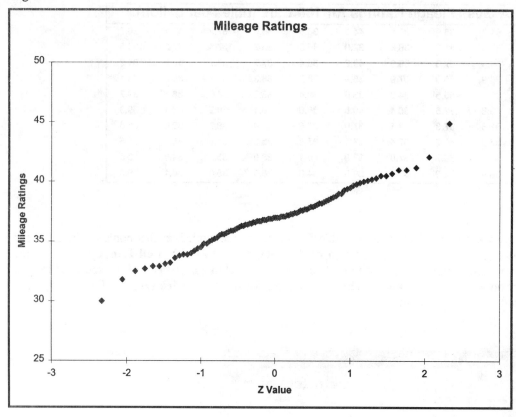

We can see from this plot that the data appear to fall in a straight line. This indicates that the data possess an approximate normal distribution.

5.4 Calculating Exponential Probabilities

To use the exponential probability tool within PHStat, **open** a new workbook and place the cursor in the upper left cell of the worksheet. **Click** on the **PHStat** menu at the top of the screen. **Select** the **Probability Distribution** option from the choices available and then select the **Exponential** option from those listed. You should open the Exponential Probability Distribution menu that looks like the one shown in Figure 5.7.

Figure 5.7

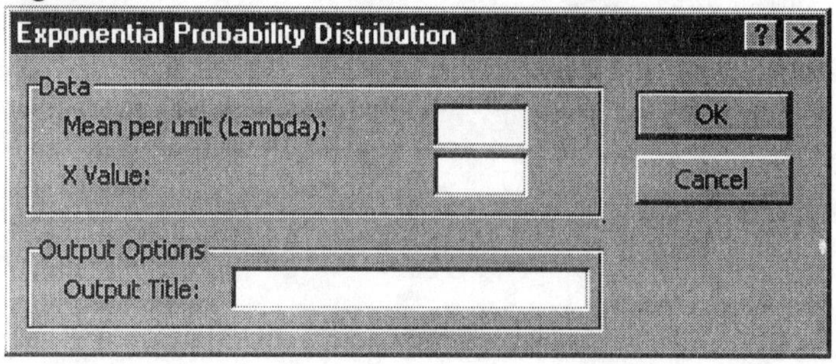

Section 5.4: Calculating Exponential Probabilities

The user is required to enter the **Value of** λ and the **X Value** of interest in the problem. An **Output Title** can be optionally selected if the user so desires. **Click OK** to finish. We illustrate with the next example.

Example 5.4: We use Example 5.13 from *Statistics for Business and Economics* found on page 248 in the text:

Suppose the length of time (in days) between sales for an automobile salesperson is modeled as an exponential distribution with λ = .5. What is the probability the salesperson goes more than 5 days without a sale?

Solution:

We utilize the **Exponential Probability Distribution** within PHStat to solve the problem. In order to solve these questions, we identify in the problem that the value of λ is λ=.5 and the **X Value** is x=5. We **enter** both of these values in the appropriate locations in the **Exponential Probability Distribution** menu shown if Figure 5.8. We **enter** an **Output Title** and finish by **clicking OK**. The output generated by PHStat is shown in Table 5.4.

Figure 5.8

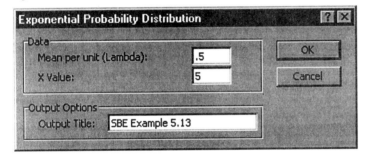

Table 5.4

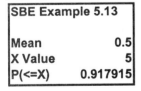

PHStat returns the probability P(x < 5) = 0.917915. To find the probability of interest, we subtract this value from 1. We get, P(x >5) = 1 - P(x < 5) = 1 - 0.917915 = 0.082085. We can verify this result using the exponential procedures discussed in the text and Table V of Appendix B. By changing the values of λ and the choice of X in the exponential probability distribution menu, the user can calculate any exponential probability desired.

Technology Lab

The following two exercises from the *Statistics for Business and Economics* text are given for you to practice the normal and exponential procedures that are available within PHStat. Included with each exercise is the PHStat output that was generated to solve each problem.

5.32 Personnel tests are designed to test a job applicant's cognitive and or physical abilities. An IQ test is an example of the former; a speed test involving the arrangement of pegs on a pegboard is an example of the latter (Cowling and James, *The Essence of Personnel Management and Industrial Relations, 1994*). A particular dexterity test is administered nationwide by a private testing service. It is known that for all tests administered last year the distribution of scores was approximately normal with mean 75 and standard deviation 7.5.
   a. A particular employer requires job candidates to score at least 80 on the dexterity test. Approximately what percentage of the test scores during the past year exceeded 80?
   b. The testing service reported to a particular employer that one of its job candidate's scores fell at the 98th percentile of the distribution (i.e., approximately 98% of the scores were lower than the candidate's, and only 2% were higher). What was the candidates score?

PHStat Output

Normal Probabilities			
Mean	75	Mean	75
Standard Deviation	7.5	Standard Deviation	7.5
		Find X and Z	
Probability for X>	80	Cumulative Percentage:	98.00%
Z Value	0.666666667	Z Value	2.053748
P(X>80)	0.252492467	X Value	90.40311

5.70 In an article published in the *European Journal of Operational Research* (Vol. 21, 1985) the vehicle-dispatching decisions of an airport-based taxi service were investigated. In modeling the system, the authors assumed travel times of successive trips to independent exponential random variables. Assume $\lambda = .05$.
   a. What is the mean trip time for the taxi service?
   b. What is the probability that a particular trip will take more than 30 minutes?
   c. Two taxis have just been dispatched. What is the probability that both will be gone for more than 30 minutes? That at least one of the taxis will return within 30 minutes?

PHStat Output

Exponential Probabilities	
Mean	0.05
X Value	30
P(<=X)	0.77686984

Chapter 6
Sampling Distributions

6.1 Introduction

Chapter 6 introduces the topic of sampling distributions to the reader. PHStat can be used when working with the sampling distribution of the sample mean covered in Section 6.3. Refer to the Normal Probability Distribution discussed in Chapter 5 of this manual for more information regarding this function. We illustrate this process in the next section.

6.2 Calculating Probabilities Using the Sampling Distribution of $\bar{x}$

The Central Limit Theorem guarantees that for large n, the sampling distribution of the sample mean possesses an approximate normal sampling distribution. In order to calculate probabilities for these sampling distributions, we must utilize the normal probability distribution within PHStat. To do this, we **open** a new workbook and place the cursor in the upper left cell of the worksheet. **Click** on the **PHStat** menu at the top of the screen. **Select** the **Probability Distribution** option from the choices available and then select the **Normal** option from those listed. You should open the Normal Probability Distribution menu that looks like the one shown in Figure 6.1.

Figure 6.1

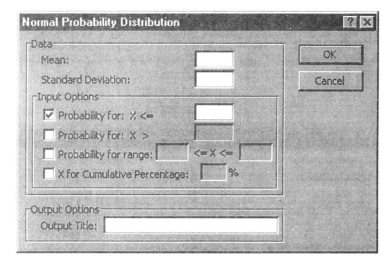

The user is required to enter the **Mean**, μ, and the **Standard Deviation**, σ, from the normal probability distribution of interest. The user then must specify the type of Input Options that they desire. Several different probability options are available as is finding a specified value of X in the normal distribution. An **Output Title** can be optionally selected if the user so desires. **Click OK** to finish. We illustrate how to use this menu when working with the sampling distribution of $\bar{x}$ in the next example.

84 Chapter 6: Sampling Distributions

Example 6.1: We use Example 6.7 from *Statistics for Business and Economics* found on pages 273-274 in the text.

Suppose we have selected a random sample of n = 25 observations from a population with mean equal to 80 and standard deviation equal to 5. It is known that the population is not extremely skewed. Find the probability that the sample mean will be larger than 82.

Solution:

The material in Chapter 6 of the text indicates that the sampling distribution of the sample mean will be approximately normal with a mean of 80 and a standard deviation of 1. We utilize this knowledge and the **Normal Probability Distribution** within PHStat to solve the problem. In order to solve this problem, we identify in the problem that the mean is $\mu=80$ and the standard deviation is $\sigma=1$. We **enter** both of these values in the appropriate locations in the **Normal Probability Distribution** menu shown if Figure 6.2. We choose here to **specify** the **Input Option Probability for: X >** as the question we wish to solve is P(x > 82). We **enter** the value **82** in the appropriate space in the menu. We also **enter** an **Output Title** and finish by **clicking OK**. The output generated by PHStat is shown in Table 6.1.

Figure 6.2

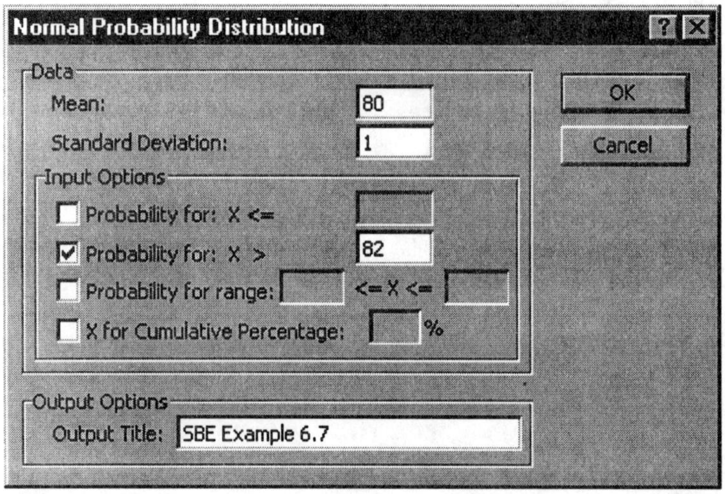

Table 6.1

SBE Example 6.7	
Mean	80
Standard Deviation	1
Probability for X>	82
Z Value	2
P(X>82)	0.022750062

We see from the output that the probability is 0.02775. The normal probability function within PHStat can be used to solve many different sampling distribution problems by changing the values of the mean, the standard deviation, and the value of X that is of interest in the problem.

Technology Lab

The following exercise from the *Statistics for Business and Economics* text is given for you to practice the normal procedure that is available within PHStat. Included with the exercise is the PHStat output that was generated to solve the problem.

6.41 Last year a company began a program to compensate its employees for unused sick days, paying each employee a bonus of one-half the usual wage earned for each unused sick day. The question that naturally arises is: "Did this policy motivate employees to use fewer sick days?" *Before* last year, the number of sick days used by employees had a distribution with a mean of 7 days and a standard deviation of 2 days.

   a. Assuming that these parameters did not change last year, find the approximate probability that the sample mean number of sick days used by 100 employees chosen at random was less than or equal to 6.4 last year.

PHStat Output

SBE Exercise 6.41	
Mean	7
Standard Deviation	0.2
Probability for X<=	6.4
Z Value	-3
P(X<=6.4)	0.001349967

Chapter 7
Inferences Based on a Single Sample:
Estimation with Confidence Intervals

7.1 Introduction

Chapter 7 introduces the reader to estimating population parameters with confidence intervals. Two parameters, the population mean and the population proportion, are studied in the chapter. The reader is also introduced to the topic of sample size determination, as it follows nicely from the estimation material presented.

PHStat provides calculation of confidence intervals for both means and proportions. There are two procedures presented for estimating a population mean; when the population standard deviation is known, and when the population standard deviation is unknown. Since the population standard deviation is almost never known, we concentrate our work on the unknown standard deviation case. The confidence intervals for a population mean can be constructed using the actual data itself or from entering summary information into the appropriate menu. We give examples of both.

The confidence interval for a population proportion requires the user to enter both the number of successes and the sample size into the PHStat menu. There is no option that allows the user to specify a data set for PHStat to use when working with proportions.

Lastly, PHStat gives a procedure for determining the sample size necessary when estimating both means and proportions. The user must enter the various pieces of information required to estimate sample sizes and PHStat calculates the sample size required.

The following examples from *Statistics for Business and Economics* are solved using PHStat in this chapter:

Excel Companion		Statistics for Business and Economics	
Example	Page	SBE	SBE Page(s)
7.1	88	Example 7.2	301-302
7.2	91	Example 7.4	312
7.3	93	Example 7.6	318-319
7.4	94	Example 7.7	320-321

7.2 Estimation of a Population Mean - Sigma Unknown

When estimating a population mean, it is highly unlikely that the population standard deviation will be known. In such cases, it is necessary to estimate the value of the population standard deviation. We illustrate how PHStat can be used if such an estimation is desired.

To use the estimation tool within PHStat, **open** a new workbook and place the cursor in the upper left cell of the worksheet. **Click** on the **PHStat** menu at the top of the screen. Select the **Confidence Intervals** option from the choices available and then select the **Estimate for the Mean, sigma unknown** option from those listed. You should open the Estimate for the Mean, sigma unknown menu that looks like the one shown in Figure 7.1.

Figure 7.1

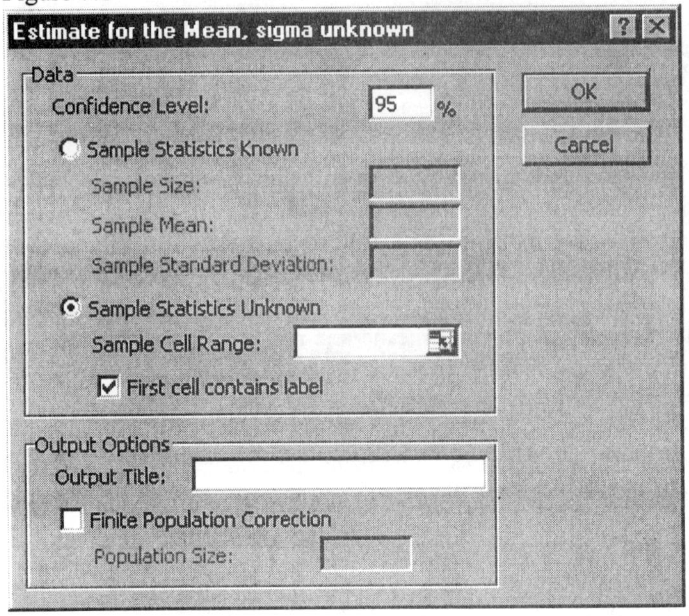

[Dialog box: Estimate for the Mean, sigma unknown
- Data:
  - Confidence Level: 95 %
  - ○ Sample Statistics Known
    - Sample Size:
    - Sample Mean:
    - Sample Standard Deviation:
  - ● Sample Statistics Unknown
    - Sample Cell Range:
    - ☑ First cell contains label
- Output Options:
  - Output Title:
  - ☐ Finite Population Correction
    - Population Size:
- OK / Cancel]

The user is required to enter the **Confidence Level** and then has the choice of how to enter the data, either using the **Sample Statistics Known** or **Sample Statistics Unknown** option. An **Output Title** can be optionally selected if the user so desires. **Click OK** to finish. We illustrate both choices of entering the data with the next example.

Example 7.1: We use Example 7.2 from *Statistics for Business and Economics* found on pages 301-302.

Some quality control experiments require destructive sampling (i.e., the test to determine whether the item is defective destroys the item) in order to measure some particular characteristic of the product. The cost of destructive sampling often dictates small samples. For example, suppose a manufacturer of printers for personal computers wishes to estimate the mean number of characters printed before the printhead fails. Suppose the printer manufacturer tests n = 15 randomly selected printheads and records the number of characters printed until failure of each. These 15 measurements (in millions of characters) are listed in Table 7.1 below. Form a 99% confidence interval for the mean number of characters printed before the printhead fails.

Table 7.1

Number of Characters (In Millions)			
1.13	1.32	1.18	1.25
1.36	1.33	0.92	1.48
1.2	1.43	1.07	1.29
1.55	0.85	1.22	

Section 7.2: Estimation of a Population Mean - Sigma Unknown

Solution:

We first solve Example 7.1 utilizing the sample statistics unknown option specified above. We **open the data set SBE Example 7-2** from the accompanying disk. We **enter** the **Confidence Level 99%** in the menu and **click** on the Sample Statistics Unknown option (see Figure 7.2). We **specify the range of the sample data** in the worksheet in the appropriate space in the menu and **add** the **title, SBE Example 7.2.** We **click OK** to finish. The output generated by PHStat is shown in Table 7.2

Table 7.2

SBE Example 7.2	
Sample Standard Deviation	0.19316413
Sample Mean	1.24
Sample Size	15
Confidence Level	99%
Standard Error of the Mean	0.049874764
Degrees of Freedom	14
t Value	2.976848918
Interval Half Width	0.148469637
Interval Lower Limit	1.09
Interval Upper Limit	1.39

Figure 7.2

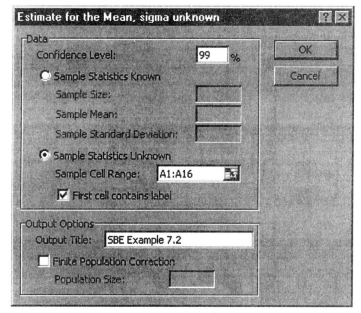

We see from the output that the confidence interval for the population mean stretches from 1.09 to 1.39. We now utilize the calculated sample mean and standard deviation found in the *Statistics for Business and Economics* text for Example 7.2 to illustrate how the estimation menu can be used when the sample statistics are known.

Solution:

We **enter** the **Confidence Level 99%** in the menu and **click** on the Sample Statistics Know option (see Figure 7.3). We **specify** the **Sample Size, Sample Mean,** and **Sample Standard Deviation** in the appropriate locations in the menu. Again, we opt to **add** the **title, SBE Example 7.2.** We **click OK** to finish. The output generated by PHStat is shown in Table 7.3.

Table 7.3

SBE Example 7.2	
Sample Standard Deviation	0.193
Sample Mean	1.239
Sample Size	15
Confidence Level	99%
Standard Error of the Mean	0.049832386
Degrees of Freedom	14
t Value	2.976848918
Interval Half Width	0.148343484
Interval Lower Limit	1.09
Interval Upper Limit	1.39

Figure 7.3

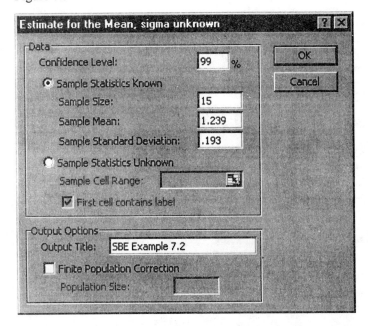

We see that both techniques lead to the same confidence interval for the population mean.

7.3 Estimation of a Population Proportion

When estimating a population proportion, it is necessary for the user to know both the sample size and the number of successes in the sample. These values must be keyed into their appropriate places in order for PHStat to generate a confidence interval for a single population proportion.

To use the estimation tool within PHStat, **open** a new workbook and place the cursor in the upper left cell of the worksheet. **Click** on the **PHStat** menu at the top of the screen. **Select** the **Confidence Intervals** option from the choices available and then select the **Estimate for the Proportion** option from those listed. You should open the Estimate for the Proportion menu that looks like the one shown in Figure 7.4.

Figure 7.4

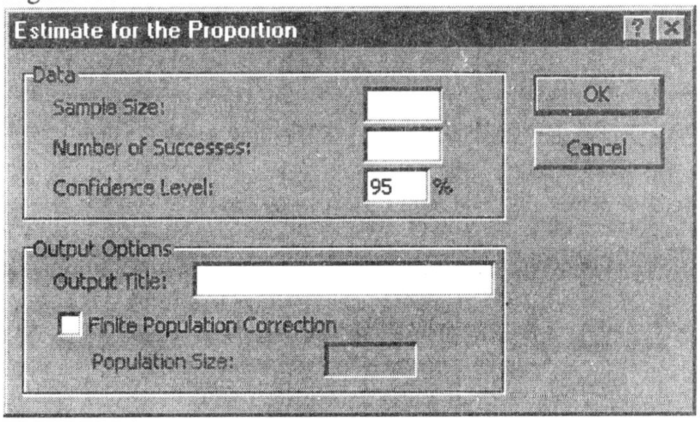

The user is required to enter the **Sample Size**, the **Number of Successes**, and the **Confidence Level**. An **Output Title** can be optionally selected if the user so desires. **Click OK** to finish. We illustrate estimating a population proportion with the next example.

Example 7.2: We use Example 7.4 from Statistics for Business and Economics found on page 312 of the text.

Many public polling agencies conduct surveys to determine the current consumer sentiment concerning the state of the economy. For example, the Bureau of Economic and Business Research (BEBR) at the University of Florida conducts quarterly surveys to gauge consumer sentiment in the Sunshine State. Suppose that BEBR randomly samples 484 consumers and finds that 257 are optimistic about the state of the economy. Use a 90% confidence interval to estimate the proportion of all consumers in Florida who are optimistic about the state of the economy.

Solution:

We first identify that the sample size is $n = 484$, the number of successes is $X = 257$ and the confidence level is 90%. We **enter** the **Sample Size, Number of Successes,** and the **Confidence Level** in the menu as shown in Figure 7.5. We **enter** the title, SBE Example 7.4. We **click OK** to finish. The output generated by PHStat is shown in Table 7.5.

Figure 7.5

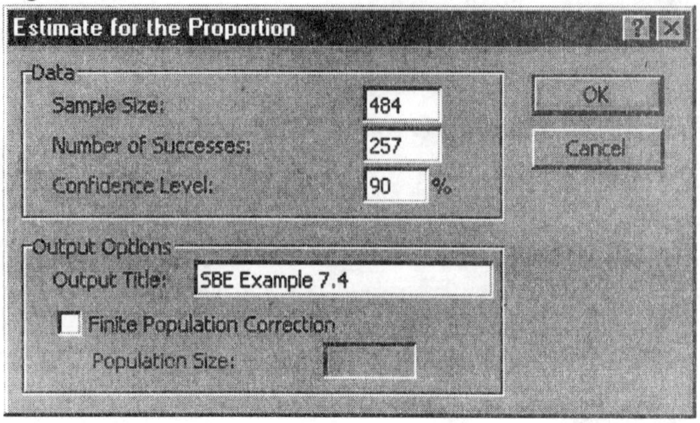

```
Estimate for the Proportion
  Data
    Sample Size:            484           OK
    Number of Successes:    257           Cancel
    Confidence Level:       90  %
  Output Options
    Output Title:  SBE Example 7.4
    ☐ Finite Population Correction
       Population Size:
```

Table 7.4

SBE Example 7.4	
Sample Size	484
Number of Successes	257
Confidence Level	90%
Sample Proportion	0.530991736
Z Value	-1.644853
Standard Error of the Proportion	0.022683572
Interval Half Width	0.037311142
Interval Lower Limit	**0.493680594**
Interval Upper Limit	**0.568302877**

We see from the output that the confidence interval for the population proportion stretches from 0.4937 to 0.5683. Confidence intervals for other population proportions can be calculated in a similar manner simply by changing the values of the sample size, number of successes, and the confidence level.

7.4 Determining the Sample Size

Section 7.4 in the *Statistics for Business and Economics* text offers the reader a technique to determine the sample size necessary when estimating both means and proportions. PHStat offers an easy method for determining the sample size. Click on the PHStat menu at the top of the screen and select the Sample Size option. You then have the choice of selecting either the Determination for the Mean or the Determination for the Proportion option. Both menus are shown below in Figure 7.6.

Figure 7.6

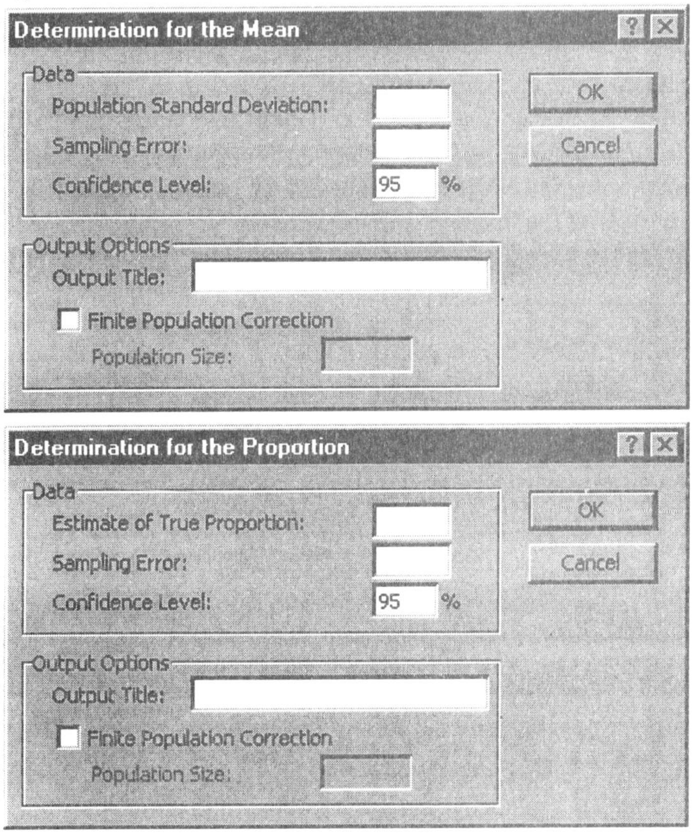

We begin by taking a look at sample size determination when estimating a population mean.

7.4.1 Determining the Sample Size for Means

The sample size determination for means menu requires the user to know the values of the **Population Standard Deviation** (or have an estimate of it), the **Sampling Error**, and the desired **Confidence Level**. These values are entered in the menu at the appropriate locations, a **Title** is added if desired, and the user **clicks** on **OK** to finish. We illustrate this procedure with an example.

Example 7.3: We use Example 7.6 from the *Statistics for Business and Economics* text found on pages 318-319.

The manufacturer of official NFL footballs uses a machine to inflate its new balls to a pressure of 13.5 pounds. When the machine is properly calibrated, the mean inflation pressure is 13.5 pounds, but uncontrollable factors cause the pressures of individual footballs to vary randomly from about

13.3 to 13.7 pounds. For quality control purposes, the manufacturer wishes to estimate the mean inflation pressure to within .025 pound of its true value with a 99% confidence. What sample size should be used?

Solution:

We note in the solution to the problem on page 319 in the text, the estimate of the population standard deviation is found using s ≈ Range/4 = .4/4 =.1. In addition, we see that the sampling error is stated at .025 and the confidence level is 99%. We **enter** these **values** in the **Determination for the Mean** menu as shown if Figure 7.7. We **specify** an **Output Title** and **click OK**. The PHStat output is shown in Table 7.5.

Figure 7.7

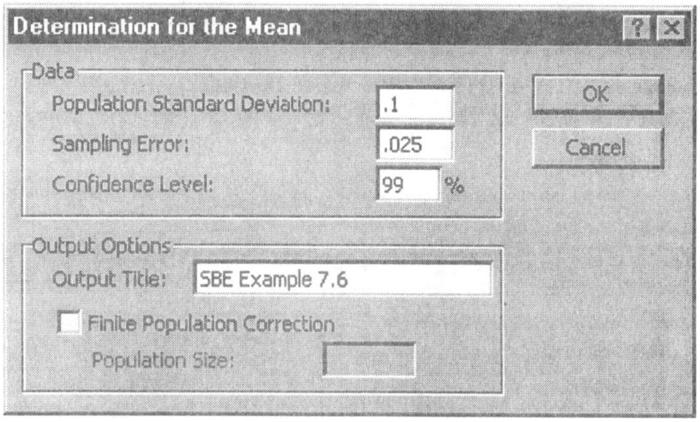

```
┌─Determination for the Mean──────────────── ? X ┐
│ ┌─Data─────────────────────────────┐          │
│ │ Population Standard Deviation:  .1    │  OK     │
│ │ Sampling Error:                 .025  │          │
│ │                                        │ Cancel  │
│ │ Confidence Level:              99  %  │          │
│ └──────────────────────────────────┘          │
│ ┌─Output Options───────────────────┐          │
│ │ Output Title:  SBE Example 7.6        │          │
│ │ ☐ Finite Population Correction        │          │
│ │ Population Size:                      │          │
│ └──────────────────────────────────┘          │
└───────────────────────────────────────────────┘
```

Table 7.5

SBE Example 7.6	
Population Standard Deviation	0.1
Sampling Error	0.025
Confidence Level	99%
Z Value	-2.57583451
Calculated Sample Size	106.1587751
Sample Size Needed	107

We find the sample size needed is n = 107. We now illustrate how to use this procedure when working with population proportions.

7.4.2 Determining the Sample Size for Proportions

The sample size determination for proportions menu requires the user to know the values of **Estimate of True Proportion**, the **Sampling Error**, and the desired **Confidence Level**. These values are entered in the menu at the appropriate locations, a **Title** is added if desired, and the user **clicks** on **OK** to finish. We illustrate this procedure with an example.

Example 7.4: We use Example 7.7 from the *Statistics for Business and Economics* text found on pages 320-321.

Section 7.4: Determining the Sample Size

A cellular telephone manufacturer that entered the postregulation market too quickly has an initial problem with excessive customer complaints and consequent returns of the cell phones for repair or replacement. The manufacturer wants to determine the magnitude of the problem in order to estimate its warranty liability. How many cellular telephones should the company randomly sample from its warehouse and check in order to estimate the fraction defective, p, to within .01 with 90% confidence?

Solution:

We note in the solution to the problem on page 320 that the estimate of the population proportion is determined to be the value .1. In addition, we see that the sampling error is stated at .01 and the confidence level is 90%. We **enter** these **values** in the **Determination for the Proportion** menu as shown if Figure 7.8. We **specify** an **Output Title** and **click OK**. The PHStat output is shown in Table 7.6.

Figure 7.8

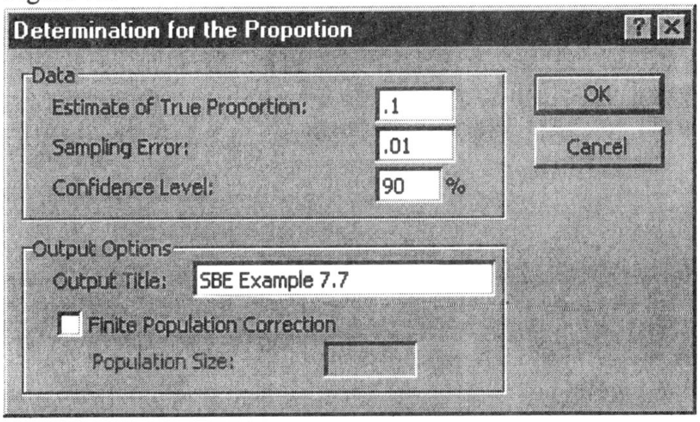

Determination for the Proportion

Data
- Estimate of True Proportion: .1
- Sampling Error: .01
- Confidence Level: 90 %

Output Options
- Output Title: SBE Example 7.7
- ☐ Finite Population Correction
- Population Size:

Table 7.6

SBE Example 7.7	
Estimate of True Proportion	0.1
Sampling Error	0.01
Confidence Level	90%
Z Value	-1.644853
Calculated Sample Size	2434.987254
Sample Size Needed	**2435**

We find the sample size needed is n = 2,435. Note that the sample size here differs slightly from the sample size found in the text due to the increase accuracy of the Z value that is used by PHStat (Z = 1.644853 vs. Z = 1.645).

Technology Lab

The following exercise from the *Statistics for Business and Economics* text is given for you to practice the normal procedure that is available within PHStat. Included with the exercise is the PHStat output that was generated to solve the problem.

7.89 A company is interested in estimating μ, the mean number of days of sick leave taken by all its employees. The firm's statistician selects at random 100 personnel files and notes the number of sick days taken by each employee. The following sample statistics are computed: $\bar{x} = 12.2$ days, $s = 10$ days.
   a. Estimate μ using a 90% confidence interval.
   b. How many personnel files would the statistician have to select in order to estimate μ to within 2 days with a 99% confidence interval?

PHStat Output

SBE Exercise 7.89 a.	
Sample Standard Deviation	10
Sample Mean	12.2
Sample Size	100
Confidence Level	90%
Standard Error of the Mean	1
Degrees of Freedom	99
t Value	1.660391717
Interval Half Width	1.660391717
Interval Lower Limit	10.54
Interval Upper Limit	13.86

SBE Exercise 7.89 b.	
Population Standard Deviation	10
Sampling Error	2
Confidence Level	99%
Z Value	-2.57583451
Calculated Sample Size	165.8730862
Sample Size Needed	166

7.94 According to the U.S. Bureau of Labor Statistics, one of every 80 American workers (i.e., 1.3%) is fired or laid off. Are employees with cancer fired or laid off at the same rate? To answer this question, *Working Women* magazine and Amgen - a company that makes drugs to lessen chemotherapy side effects - conducted a telephone survey of 100 cancer survivors who worked while undergoing treatment (*Tampa Tribune*, Sept. 25, 1996). Of these 100 cancer patients, 7 were fired or laid off due to their illness.

   a. Construct a 90% confidence interval for the true percentage of all cancer patients who are fired or laid off due to their illness.

PHStat Output

SBE Exercise 7.94 a.	
Sample Size	100
Number of Successes	7
Confidence Level	90%
Sample Proportion	0.07
Z Value	-1.644853
Standard Error of the Proportion	0.025514702
Interval Half Width	0.041967934
Interval Lower Limit	0.028032066
Interval Upper Limit	0.111967934

7.57   A gigantic warehouse located in Tampa, Florida, stores approximately 60 million empty aluminum beer and soda cans. Recently, a fire occurred at the warehouse. The smoke from the fire contaminated many of the cans with blackspot, rendering them unusable. A University of South Florida statistician was hired by the insurance company to estimate the true proportion of cans in the warehouse that were contaminated by the fire. How many aluminum cans should be randomly sampled to estimate the true proportion to within .02 with 90% confidence?

PHStat Output

SBE Exercise 7.57	
Estimate of True Proportion	0.5
Sampling Error	0.02
Confidence Level	90%
Z Value	-1.644853
Calculated Sample Size	1690.963371
Sample Size Needed	1691

Chapter 8
Inferences Based on a Single Sample: Tests of Hypothesis

8.1 Introduction

Chapter 8 introduces the reader to the concepts of hypothesis testing. The general theory and concepts of the test of hypothesis are then examined for inferences based on a single sample. Tests for both a single population mean and a single population proportion are discussed in Chapter 8. In addition, the observed significance level of a test of hypothesis is explained and demonstrated in several examples.

PHStat provides calculation of tests of hypothesis for both means and proportions. There are two procedures presented for testing a population mean; when the population standard deviation is known, and when the population standard deviation is unknown. Since the population standard deviation is almost never known, we concentrate our work on the unknown standard deviation case. The test of hypothesis for a population mean can be constructed using the actual data itself or from entering summary information into the appropriate menu. We give examples of both.

The test of hypothesis for a population proportion requires the user to enter both the number of successes and the sample size into the PHStat menu. There is no option that allows the user to specify a data set for PHStat to use when working with proportions.

The following examples from *Statistics for Business and Economics* are solved using PHStat in this chapter:

Excel Companion		Statistics for Business and Economics	
Example	Page	SBE	SBE Page(s)
8.1	100	Example 8.5	362-363
8.2	104	Example 8.7	370-371

8.2 Tests of Hypothesis of a Population Mean - Sigma Unknown

When testing a population mean, it is highly unlikely that the population standard deviation will be known. In such cases, it is necessary to estimate the value of the population standard deviation. We illustrate how PHStat can be used if such a test of hypothesis is desired.

To use the test of hypothesis tool within PHStat, **open** a new workbook and place the cursor in the upper left cell of the worksheet. **Click** on the **PHStat** menu at the top of the screen. **Select** the**One-Sample Tests** option from the choices available and then select the **t Test for the Mean, sigma unknown** option from those listed. You should open the t Test for the Mean, sigma unknown menu that looks like the one shown in Figure 8.1.

100   Chapter 8: Inferences Based on a Single Sample: Tests of Hypothesis

Figure 8.1

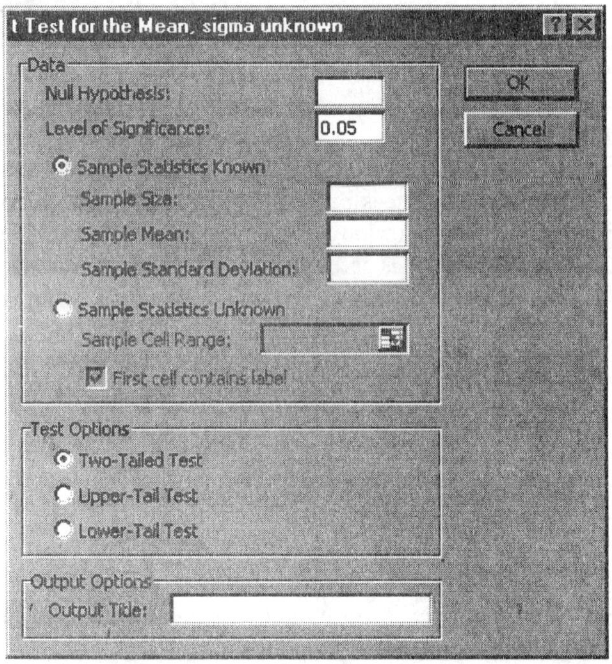

[Dialog box: t Test for the Mean, sigma unknown — Data: Null Hypothesis, Level of Significance: 0.05, Sample Statistics Known (Sample Size, Sample Mean, Sample Standard Deviation), Sample Statistics Unknown (Sample Cell Range, First cell contains label); Test Options: Two-Tailed Test, Upper-Tail Test, Lower-Tail Test; Output Options: Output Title; OK, Cancel]

The user is required to enter the **Null Hypothesis**, a **Level of Significance**, and the direction of the test in the **Test Option**. The user then has the choice of how to enter the data, either using the **Sample Statistics Known** or **Sample Statistics Unknown** option. An **Output Title** can be optionally selected if the user so desires. **Click OK** to finish. We illustrate both choices of entering the data with the next example.

Example 8.1:   We use Example 8.5 found on pages 362-363 of the *Statistics for Business and Economics* text.

A major car manufacturer wants to test a new engine to determine whether it meets new air pollution standards. The mean emission μ of all engines of this type must be less than 20 parts per million of carbon. Ten engines are manufactured for testing purposes, and the emission level of each is determined. The data are shown in Table 8.1 below. Do the data supply sufficient evidence to allow the manufacturer to conclude that this type of engine meets the pollution standard? Assume that the production process is stable and the manufacturer is willing to risk a Type I error with probability $\alpha = .01$.

Table 8.1

Emission Level of Engine
15.6
16.2
22.5
20.5
16.4
19.4
16.6
17.9
12.7
13.9

Section 8.2: Tests of Hypothesis of a Population Mean - Sigma Unknown

Solution:

We first solve example 8.1 utilizing the sample statistics unknown option specified above. We **open** the data set **SBE Example 8-5** from the accompanying disk. We **enter** the **Null Hypothesis** value of 20 parts per million in the menu, enter a .01 Level of Significance, and **select** the **Lower-Tail Test** option. We then **click** on the Sample Statistics Unknown option (see Figure 8.2). We **specify the range of the sample data** in the worksheet in the appropriate space in the menu and **add** the **title, SBE Example 8.5**. We **click OK** to finish. The output generated by PHStat is shown in Table 8.2

Figure 8.2

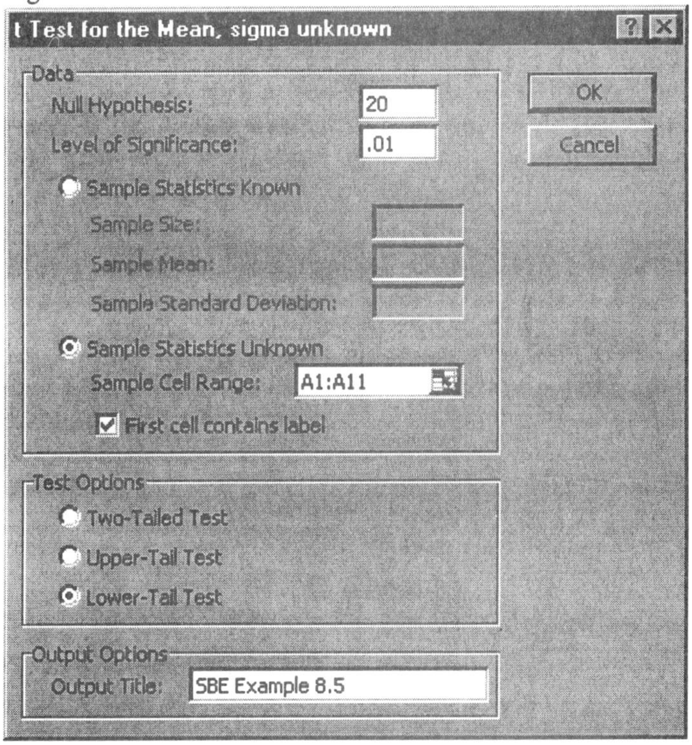

Table 8.2

SBE Example 8-5		
Null Hypothesis	$\mu=$	20
Level of Significance		0.01
Sample Size		10
Sample Mean		17.17
Sample Standard Deviation		2.981442604
Standard Error of the Mean		0.942814934
Degrees of Freedom		9
t Test Statistic		-3.001649526
Lower-Tail Test		
Lower Critical Value		-2.821434464
p-Value		0.007458207
Reject the null hypothesis		

Chapter 8: Inferences Based on a Single Sample: Tests of Hypothesis

We see in the printout that the test statistic is reported to be t = -3.002 and the p-value for the test is reported to be p = 0.00746. Both of these values indicate that the null hypothesis should be rejected at $\alpha = .01$. We now utilize the calculated sample mean and standard deviation found in the printout above for Example 8.5 to illustrate how the test of hypothesis menu can be used when the sample statistics are known.

Solution:

We **enter** the **Null Hypothesis** value of **20** parts per million, the **.01 Level of Significance**, and **the Lower-Tail Test** just like we did in the last example. This time, however, we **specify** the **Sample Size, Sample Mean,** and **Sample Standard Deviation** in the appropriate locations in the **Sample Statistics Known** menu (see Figure 8.3). Again, we opt to **add** the **title, SBE Example 8.5.** We click **OK** to finish. The output generated by PHStat is shown in Table 8.3.

Table 8.3

SBE Example 8-5	
Null Hypothesis µ=	20
Level of Significance	0.01
Sample Size	10
Sample Mean	17.17
Sample Standard Deviation	2.981442604
Standard Error of the Mean	0.942814934
Degrees of Freedom	9
t Test Statistic	-3.001649526
Lower-Tail Test	
Lower Critical Value	-2.821434464
p-Value	0.007458207
Reject the null hypothesis	

Figure 8.3

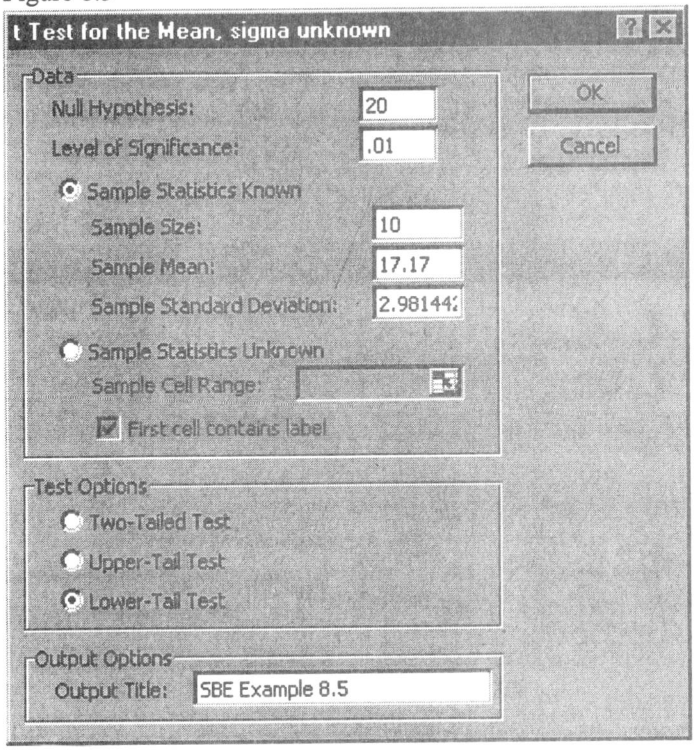

```
t Test for the Mean, sigma unknown                    [?][X]
┌─Data──────────────────────────────────┐
│  Null Hypothesis:          [ 20      ]     [   OK    ]  │
│  Level of Significance:    [ .01     ]     [ Cancel  ]  │
│  ⊙ Sample Statistics Known                              │
│     Sample Size:           [ 10      ]                  │
│     Sample Mean:           [ 17.17   ]                  │
│     Sample Standard Deviation: [2.98144]                │
│  ○ Sample Statistics Unknown                            │
│     Sample Cell Range:     [         ]                  │
│     ☑ First cell contains label                         │
└─────────────────────────────────────────┘
┌─Test Options──────────────────────────┐
│  ○ Two-Tailed Test                     │
│  ○ Upper-Tail Test                     │
│  ⊙ Lower-Tail Test                     │
└─────────────────────────────────────────┘
┌─Output Options────────────────────────┐
│  Output Title: [ SBE Example 8.5     ] │
└─────────────────────────────────────────┘
```

We see that both techniques lead to the same test of hypothesis results when testing the population mean.

8.3 Tests of Hypothesis of a Population Proportion

When testing a population proportion, it is necessary for the user to know both the sample size and the number of successes in the sample. These values must be keyed into their appropriate places in order for PHStat to generate a confidence interval for a single population proportion.

To use the estimation tool within PHStat, **open** a new workbook and place the cursor in the upper left cell of the worksheet. **Click** on the **PHStat** menu at the top of the screen. **Select** the **One-Sample Tests** option from the choices available and then select the **Z Test for the Proportion** option from those listed. You should open the Z Test for the Proportion menu that looks like the one shown in Figure 8.4.

104 Chapter 8: Inferences Based on a Single Sample: Tests of Hypothesis

Figure 8.4

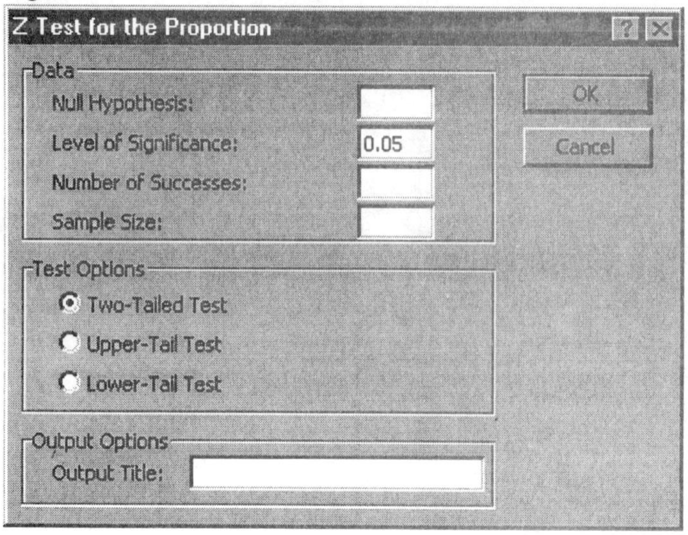

```
Z Test for the Proportion
┌─Data─────────────────────────┐
│  Null Hypothesis:     [      ]       [   OK   ]
│  Level of Significance: [0.05]       [ Cancel ]
│  Number of Successes: [      ]
│  Sample Size:         [      ]
├─Test Options────────────────┤
│  ● Two-Tailed Test
│  ○ Upper-Tail Test
│  ○ Lower-Tail Test
├─Output Options──────────────┤
│  Output Title: [                    ]
└─────────────────────────────┘
```

The user is required to enter the **Null Hypothesis** value, the **Level of Significance,** the **Number of Successes,** the **Sample Size,** and the direction of the test to be conducted in the **Test Options** area of the menu. An **Output Title** can be optionally selected if the user so desires. **Click OK** to finish. We illustrate estimating a population proportion with the next example.

Example 8.2: We use Example 8.7 from *Statistics for Business and Economics* found on pages 370-371 of the text.

The reputations (and hence sales) of many businesses can be severely damaged by shipments of manufactured items that contain a large percentage of defectives. For example, a manufacturer of alkaline batteries may want to be reasonably certain that fewer than 5% of its batteries are defective. Suppose 300 batteries are randomly selected from a very large shipment; each is tested and 10 defective batteries are found. Does this provide sufficient evidence for the manufacturer to conclude that the fraction defective in the entire shipment is less than .05? Use $\alpha = .01$.

Solution:

We first identify that the sample size is $n = 300$, the number of successes is $X = 10$, and the level of significance is .01. We want to use the null hypothesis value of .05 and conduct a lower-tail test. We enter the **Null Hypothesis** value of **.05**, the **Level of Significance** of **.01**, the **Number of Successes (10)**, and the **Sample Size (300)** in their appropriate locations on the menu (see Figure 8.5). We then select the **Lower-Tail Test** from the Test Options area of the menu. We enter the title, SBE Example 8.7. We click OK to finish. The output generated by PHStat is shown in Table 8.4.

Section 8.3: Tests of Hypothesis of a Population Proportion

Figure 8.5

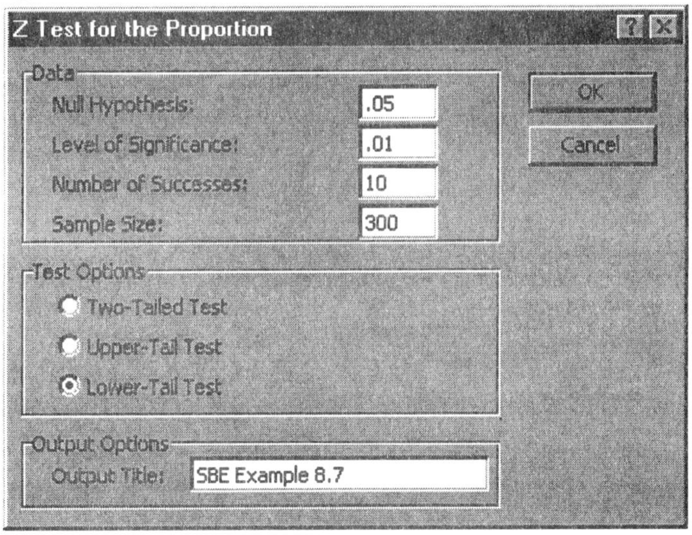

```
Z Test for the Proportion
  Data
    Null Hypothesis:          .05           [ OK ]
    Level of Significance:    .01           [ Cancel ]
    Number of Successes:      10
    Sample Size:              300
  Test Options
    ( ) Two-Tailed Test
    ( ) Upper-Tail Test
    (•) Lower-Tail Test
  Output Options
    Output Title: SBE Example 8.7
```

Table 8.4

SBE Example 8.7	
Null Hypothesis $p=$	0.05
Level of Significance	0.01
Number of Successes	10
Sample Size	300
Sample Proportion	0.033333333
Standard Error	0.012583057
Z Test Statistic	-1.324532357
Lower-Tail Test	
Lower Critical Value	-2.326341928
p-Value	0.092663215
Do not reject the null hypothesis	

We see from the output that the test of hypothesis cannot be rejected when testing at $\alpha = .01$. There is insufficient evidence to indicate that the population proportion of defective batteries is less than 5. Tests of hypothesis for other population proportions can be conducted in a similar manner simply by changing the values of the sample size, number of successes, the level of significance, the null hypothesis and the direction of the test.

Technology Lab

The following exercises from the *Statistics for Business and Economics* text are given for you to practice the procedures covered in the text that are available within PHStat. Included with the exercises are the PHStat outputs that were generated to solve the problems.

8.101 The trade publication *Potentials in Marketing* (Nov./Dec. 1995) surveyed its readers concerning their opinions of electronic marketing (i.e., marketing via the Internet, e-mails, CD-ROMS, etc.). A questionnaire was faxed to 1,500 randomly selected U.S. readers in August and September 1995. Of the 195 questionnaires that were returned, 37 reported that their company already had a World Wide Web site and 59 indicated that their company had plans to create one.

    a. Do these data provide sufficient evidence to reject the claim by a producer of a well-known Web browser that "more than 25% of all U.S. businesses will have web sites by the middle of 1995"?

PHStat Output

SBE Exercise 8.101	
Null Hypothesis $p=$	0.25
Level of Significance	0.05
Number of Successes	37
Sample Size	195
Sample Proportion	0.18974359
Standard Error	0.031008684
Z Test Statistic	-1.943210842
Lower-Tail Test	
Lower Critical Value	-1.644853
p-Value	0.025995275
Reject the null hypothesis	

8.108 One way of evaluating a measuring instrument is to repeatedly measure the same item and compare the average of these measurements to the item's known measured value. The difference is used to assess the instrument's accuracy (*Quality Progress*, Jan. 1993). To evaluate a particular Metlar scale, an item whose weight is known to be 16.01 ounces is weighed five times by the same operator. The measurements, in ounces, are as follows:
15.99   16.00   15.97   16.01   15.96

    a. In a statistical sense, does the average measurement differ from 16.01? Conduct the appropriate hypothesis test. What does your analysis suggest about the accuracy of the instrument?

PHStat Output

SBE Exercise 8.108		
Null Hypothesis $\mu=$		16.01
Level of Significance		0.05
Sample Size		5
Sample Mean		15.986
Sample Standard Deviation		0.020736441
Standard Error of the Mean		0.009273618
Degrees of Freedom		4
t Test Statistic		-2.587986557
	Two-Tailed Test	
Lower Critical Value		-2.776450856
Upper Critical Value		2.776450856
p-Value		0.060812856
	Do not reject the null hypothesis	

Chapter 9
Inferences Based on Two Samples:
Confidence Intervals and Tests of Hypothesis

9.1 Introduction

Chapter 9 introduces the reader to two sample problems using both the estimation and test of hypothesis techniques discussed in Chapters 7 and 8. Three types of parameters, population means, population proportions, and population variances, are studied in the chapter. The reader is also introduced to the topic of sample size determination, as it follows very nicely from the estimation material presented.

PHStat provides calculation of tests of hypotheses for comparing population means, population proportions, and population variances. PHStat provides only an independent comparison of population means. No paired difference analysis currently exists in the program. PHStat also does not provide any tool for calculating the corresponding confidence intervals for these parameters. Finally, no sample size determination techniques are available for any of the parameters presented in Chapter 9.

Each of the test of hypothesis techniques presented in PHStat require the user to have available the summary statistics of the data. Whether the user is working with means, proportions, or variances, these statistics are entered into menus to generate the desired test of hypothesis output. The following examples from *Statistics for Business and Economics* are solved using PHStat in this chapter:

Excel Companion		Statistics for Business and Economics	
Example	Page	SBE	Page(s)
9.1	110	Example 9.2	397
9.2	113	Example 9.6	426-428
9.3	115	Example 9.10	437-438

9.2 Tests For Differences in Two Means

The *Statistics for Business and Economics* text offers two techniques for comparing two population means, the independent sampling and matched pairs sampling techniques. PHStat allows the user to perform a test of hypothesis when the data has been collected using two random, independent samples. To use the test of hypothesis tool within PHStat, **open** a new workbook and place the cursor in the upper left cell of the worksheet. **Click** on the **PHStat** menu at the top of the screen. **Select** the **Two-Sample Tests** option from the choices available and then select the **t Test for Differences in Two Means** option from those listed. You should open the t Test for differences in Two Means menu that looks like the one shown in Figure 9.1.

The user is required to enter the **Hypothesized Difference**, a Level of Significance, the **Summary Statistics** from both the samples collected, and the direction of the test in the **Test Option**. An **Output Title** can be optionally selected if the user so desires. **Click OK** to finish. We illustrate the test with the next example.

110 Chapter 9: Inferences Based on a Two Samples

Figure 9.1

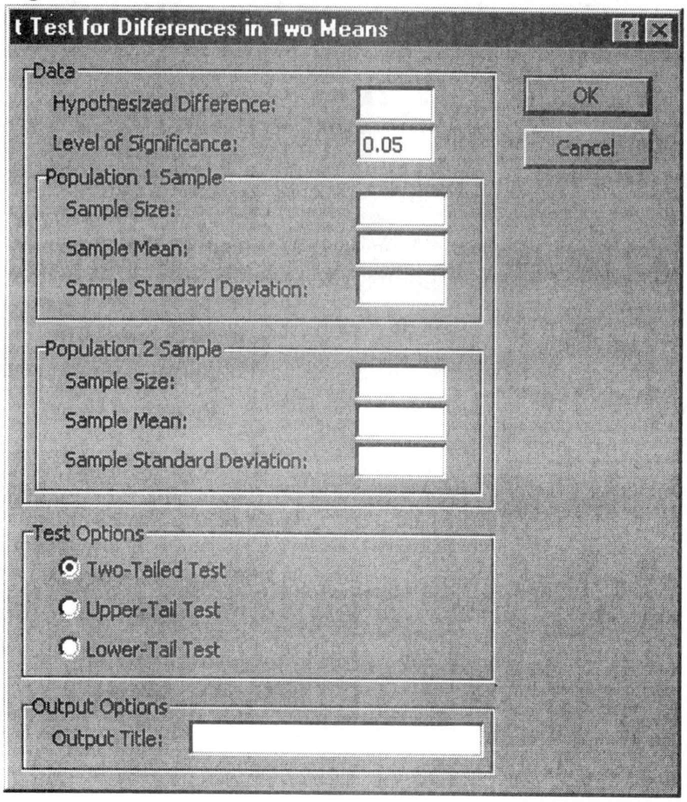

```
t Test for Differences in Two Means                    [?][X]
┌Data─────────────────────────────────┐
│  Hypothesized Difference:  [        ]      [  OK  ]  │
│  Level of Significance:    [ 0.05   ]      [ Cancel ]│
│ ┌Population 1 Sample────────────────┐│
│ │  Sample Size:              [      ]││
│ │  Sample Mean:              [      ]││
│ │  Sample Standard Deviation:[      ]││
│ └──────────────────────────────────┘│
│ ┌Population 2 Sample────────────────┐│
│ │  Sample Size:              [      ]││
│ │  Sample Mean:              [      ]││
│ │  Sample Standard Deviation:[      ]││
│ └──────────────────────────────────┘│
├Test Options────────────────────────┤
│  ⊙ Two-Tailed Test                  │
│  ○ Upper-Tail Test                  │
│  ○ Lower-Tail Test                  │
├Output Options──────────────────────┤
│  Output Title: [                  ] │
└────────────────────────────────────┘
```

Example 9.1:   We use Example 9.2 found on page 397 of the *Statistics for Business and Economics* text.

In recent years, the United States and Japan have engaged in intense negotiations regarding restrictions on trade between the two countries. One of the claims made repeatedly by U.S. officials is that many Japanese manufacturers price their goods higher in Japan than in the United States, in effect subsidizing low prices in the United States by extremely high prices in Japan. According to the U.S. argument, Japan accomplishes this by keeping competitive U.S. goods from reaching the Japanese marketplace.

An economist decided to test the hypothesis that higher retail prices are being charged for Japanese automobiles in Japan than in the United States. She obtained random samples of 50 retail sales in the United States and 30 retail sales in Japan over the same time period and for the same model of automobile, converted the Japanese sales prices from yen to dollars using current conversion rates, and obtained the summary information shown in Table 9.1. Compare the mean retail prices for the two countries using a test of hypothesis. Use $\alpha = .5$.

Table 9.1

	U.S. Sales	Japan Sales
Sample Size	50	30
Sample Mean	$16,545	$17,243
Sample Standard Deviation	$ 1,989	$ 1,843

Section 9.2: Tests for Differences in Two Means

Solution:

We **specify a Hypothesized Difference** value of 0 (because we are comparing the two means to determine if they differ), a **.05 Level of Significance**, and the **Lower-Tail Test** (to determine if the Japanese mean is higher than the U.S. mean) option. We **specify the Sample Size, Sample Mean, and Sample Standard Deviation** for both samples in the appropriate locations in the **t Test for Differences in Two Means** menu (see Figure 9.2). We opt to **add the title, SBE Example 9.2.** We click OK to finish. The output generated by PHStat is shown in Table 9.2.

Figure 9.2

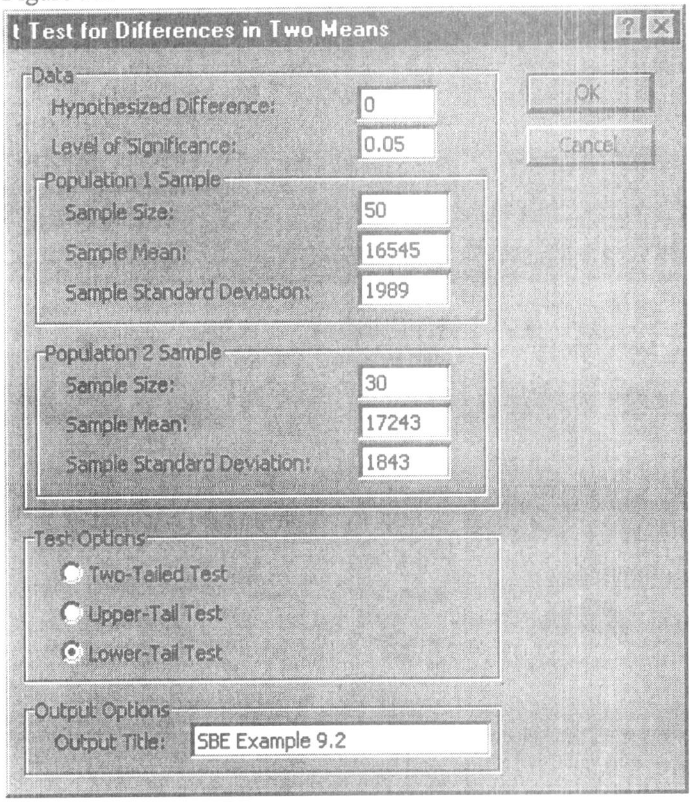

t Test for Differences in Two Means

Data:
- Hypothesized Difference: 0
- Level of Significance: 0.05

Population 1 Sample:
- Sample Size: 50
- Sample Mean: 16545
- Sample Standard Deviation: 1989

Population 2 Sample:
- Sample Size: 30
- Sample Mean: 17243
- Sample Standard Deviation: 1843

Test Options:
- ○ Two-Tailed Test
- ○ Upper-Tail Test
- ● Lower-Tail Test

Output Options:
- Output Title: SBE Example 9.2

112 Chapter 9: Inferences Based on a Two Samples

Table 9.2

SBE Example 9.2	
Hypothesized Difference	0
Level of Significance	0.05
Population 1 Sample	
Sample Mean	16545
Sample Size	50
Sample Standard Deviation	1989
Population 2 Sample	
Sample Mean	17243
Sample Size	30
Sample Standard Deviation	1843
Population 1 Sample Degrees of Freedom	49
Population 2 Sample Degrees of Freedom	29
Total Degrees of Freedom	78
Pooled Variance	3748112
Difference in Sample Means	-698
t-Test Statistic	-1.56117
Lower-Tail Test	
Lower Critical Value	-1.66463
p-Value	0.061266
Do not reject the null hypothesis	

Both the test statistic and p-value generated on the printout result in the fail to reject H_o conclusion. By changing the summary statistics above, tests can be performed for any independent samples comparison of population means.

9.3 Tests For Differences in Two Proportions

The *Statistics for Business and Economics* text describes the technique for comparing two population proportions. PHStat allows the user to perform a test of hypothesis when comparing two proportions. To use the test of hypothesis tool within PHStat, **open** a new workbook and place the cursor in the upper left cell of the worksheet. **Click** on the **PHStat** menu at the top of the screen. **Select** the **Two-Sample Tests** option from the choices available and then select the **Z Test for Differences in Two Proportions** option from those listed. You should open the Z Test for differences in Two Proportions menu that looks like the one shown in Figure 9.3.

The user is required to enter the **Hypothesized Difference**, a **Level of Significance**, the **Sample Size** and **Number of Successes** from both the samples collected, and the direction of the test in the **Test Option**. An **Output Title** can be optionally selected if the user so desires. **Click OK** to finish. We illustrate the test with the next example.

Section 9.3: Tests for Differences in Two Proportions

Figure 9.3

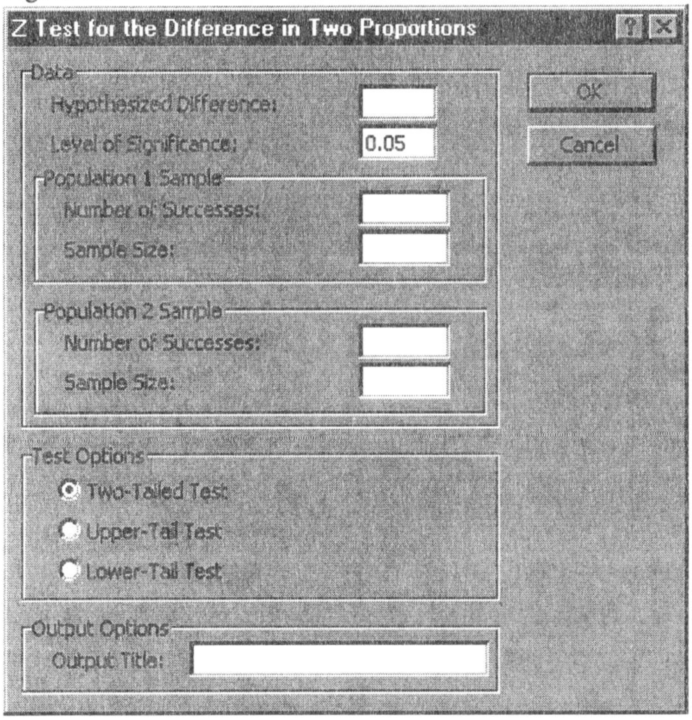

Example 9.2: We use Example 9.6 found on pages 426-428 of the *Statistics for Business and Economics* text.

A consumer advocacy group wants to determine whether there is a difference between the proportions of the two leading automobile models that need major repairs (more than $500) within two years of their purchase. A sample of 400 two-year owners of model 1 is contacted, and a sample of 500 two-year owners of model 2 is contacted. The numbers x_1 and x_2 of owners who report that their cars needed major repairs within the first two years are 53 and 78, respectively. Test the null hypothesis that no difference exists between the proportions in populations 1 and 2 needing major repairs against the alternative that a difference does exist. Use $\alpha = .10$.

Solution:

We **enter** the **Hypothesized Difference** value of **0** (because we are comparing the two proportions to determine if they differ, the **.10 Level of Significance**, and **the Two-Tail Test** (to determine if a difference exists) option. We **specify** the **Number of Successes** and the **Sample** for both samples in the appropriate locations in the **Z Test for Differences in Two Proportions** menu (see Figure 9.4). We opt to **add** the **title, SBE Example 9.6.** We **click OK** to finish. The output generated by PHStat is shown in Table 9.3.

Figure 9.4

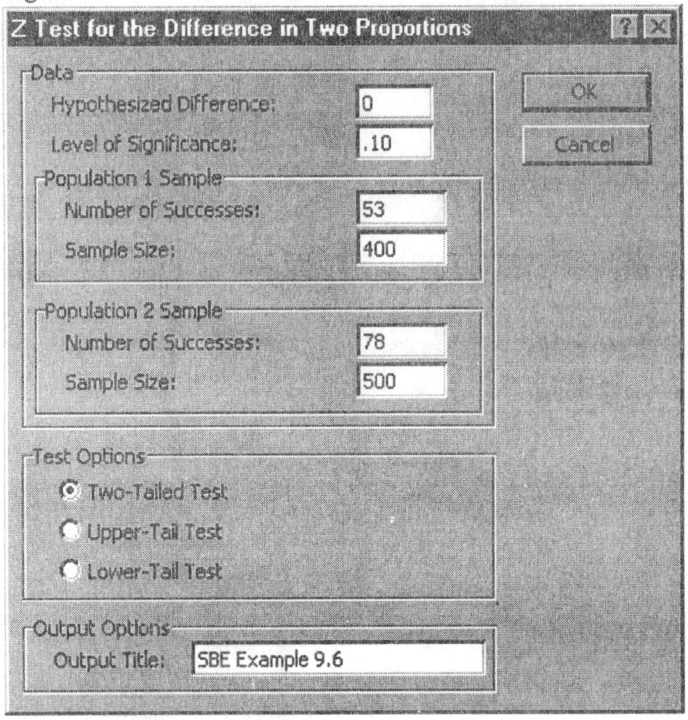

Z Test for the Difference in Two Proportions

Data:
- Hypothesized Difference: 0
- Level of Significance: .10

Population 1 Sample:
- Number of Successes: 53
- Sample Size: 400

Population 2 Sample:
- Number of Successes: 78
- Sample Size: 500

Test Options:
- ● Two-Tailed Test
- ○ Upper-Tail Test
- ○ Lower-Tail Test

Output Options:
- Output Title: SBE Example 9.6

Table 9.3

SBE Example 9.6	
Hypothesized Difference	0
Level of Significance	0.1
Group 1	
Number of Successes	53
Sample Size	400
Group 2	
Number of Successes	78
Sample Size	500
Group 1 Proportion	0.1325
Group 2 Proportion	0.156
Difference in Two Proportions	-0.0235
Average Proportion	0.145555556
Z Test Statistic	-0.993356864
Two-Tailed Test	
Lower Critical Value	-1.644853
Upper Critical Value	1.644853
p-Value	0.320536082
Do not reject the null hypothesis	

Both the test statistic and p-value generated on the printout result in the fail to reject H_o conclusion. By changing the summary statistics above, tests can be performed for any comparison of population proportions.

9.4 Tests For Differences in Two Variances

The *Statistics for Business and Economics* text describes the technique for comparing two population variances. PHStat allows the user to perform a test of hypothesis when comparing two variances. To use the test of hypothesis tool within PHStat, **open** a new workbook and place the cursor in the upper left cell of the worksheet. **Click** on the **PHStat** menu at the top of the screen. **Select** the **Two-Sample Tests** option from the choices available and then select the **F Test for Differences in Two Variances** option from those listed. You should open the F Test for differences in Two Variances menu that looks like the one shown in Figure 9.5.

The user is required to enter the **Level of Significance**, the **Sample Size** and the **Sample Standard Deviation** from both the samples collected, and the direction of the test in the **Test Option**. An **Output Title** can be optionally selected if the user so desires. **Click OK** to finish. We illustrate the test with the next example.

Figure 9.5

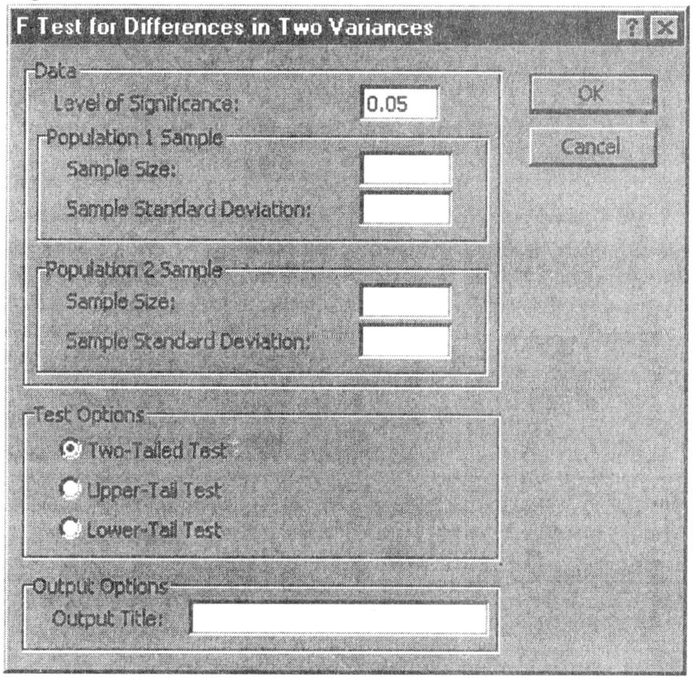

Example 9.3: We use Example 9.10 found on pages 437-438 of the *Statistics for Business and Economics* text.

A manufacturer of paper products wants to compare the variation in daily production levels at two paper mills. Independent random samples of days are selected from each mill and the production levels (in units) recorded. The following summary information was obtained:

	Mill 1	Mill 2
n	13	18
Mean	26.3	19.7
Std. Dev.	8.2	4.7

Do the data provide sufficient evidence to indicate a difference in the variability of production levels at the two paper mills? (Use $\alpha = .10$).

Chapter 9: Inferences Based on a Two Samples

Solution:

We **enter** the **.10 Level of Significance**, and **the Two-Tail Test** (to determine if a difference exists) option. We **specify** the **Sample Size** and the **Standard deviation** for both samples in the appropriate locations in the **F Test for Differences in Two Variances** menu (see Figure 9.6). We opt to **add** the title, **SBE Example 9.10.** We **click OK** to finish. The output generated by PHStat is shown in Table 9.4.

Figure 9.6

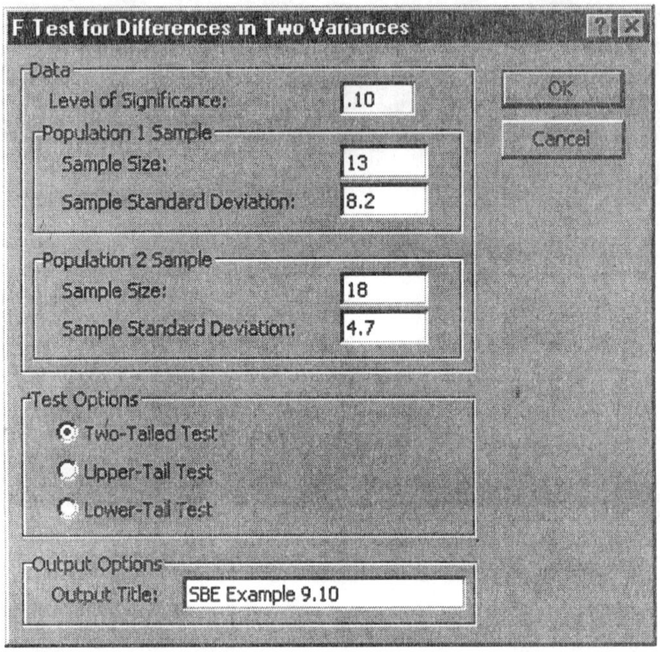

[F Test for Differences in Two Variances dialog box]
- Level of Significance: .10
- Population 1 Sample:
  - Sample Size: 13
  - Sample Standard Deviation: 8.2
- Population 2 Sample:
  - Sample Size: 18
  - Sample Standard Deviation: 4.7
- Test Options: Two-Tailed Test (selected), Upper-Tail Test, Lower-Tail Test
- Output Title: SBE Example 9.10

Table 9.4

SBE Example 9.10	
Level of Significance	0.1
Population 1 Sample	
Sample Size	13
Sample Standard Deviation	8.2
Population 2 Sample	
Sample Size	18
Sample Standard Deviation	4.7
F-Test Statistic	3.043911
Population 1 Sample Degrees of Freedom	12
Population 2 Sample Degrees of Freedom	17
Two-Tailed Test	
Lower Critical Value	0.387171
Upper Critical Value	2.380652
p-Value	0.036042
Reject the null hypothesis	

Both the test statistic and p-value generated on the printout result in the reject H_o conclusion. By changing the summary statistics above, tests can be performed for any comparison of population variances.

Technology Lab

The following exercises from the *Statistics for Business and Economics* text are given for you to practice the procedures covered in the text that are available within PHStat. Included with the exercises are the PHStat outputs that were generated to solve the problems.

9.19 One way corporations raise money for expansion is to issue *bonds*, loan agreements to repay the purchaser a specified amount with a fixed rate of interest paid periodically over the life of the bond. The sale of bonds is usually handled by an underwriting firm. Does it pay for companies to shop around for an underwriter? The reason for the question is that the price of a bond may rise or fall after is issuance. Therefore, whether a corporation receives the market price for a bond depends on the skill of its underwriter (Radcliffe, 1994). The mean change in the prices of 27 bonds handled over a 12-month period by one underwriter and in the prices of 23 bonds handled by another are shown

	Underwriter 1	Underwriter 2
Sample size	27	23
Sample Mean	-0.0491	-0.0307
Sample Variance	0.0098	0.002465

a. Do the data provide sufficient evidence to indicate a difference in the mean change in bond prices handled by the two underwriters? Test using $\alpha = .05$

PHStat Output

SBE Exercise 9.19	
Hypothesized Difference	0
Level of Significance	0.05
Population 1 Sample	
Sample Mean	-0.0491
Sample Size	27
Sample Standard Deviation	0.098995
Population 2 Sample	
Sample Mean	-0.0307
Sample Size	23
Sample Standard Deviation	0.049649
Population 1 Sample Degrees of Freedom	26
Population 2 Sample Degrees of Freedom	22
Total Degrees of Freedom	48
Pooled Variance	0.006438
Difference in Sample Means	-0.0184
t-Test Statistic	-0.80816
Two-Tailed Test	
Lower Critical Value	-2.01063
Upper Critical Value	2.010634
p-Value	0.422984
Do not reject the null hypothesis	

9.53 *Industrial Marketing Management* (Vol. 25, 1996) published a study that examined the demographics, decision-making roles, and time demands of product managers. Independent samples of $n_1 = 93$ consumer/commercial product managers and $n_2 = 212$ industrial product managers took part in the study. In the consumer/commercial group, 40% of the product managers are 40 years of age or older; in the industrial group, 54% are 40 or more years old. Make an inference about the difference between the true proportions of consumer/commercial and industrial product managers who are at least 40 years old. Justify your choice of method (confidence interval or hypothesis test) and α level. Do industrial product managers tend to be older than consumer/commercial product managers?

PHStat Output

SBE Exercise 9.53	
Hypothesized Difference	0
Level of Significance	0.05
Group 1	
Number of Successes	37
Sample Size	93
Group 2	
Number of Successes	114
Sample Size	212
Group 1 Proportion	0.397849462
Group 2 Proportion	0.537735849
Difference in Two Proportions	-0.139886387
Average Proportion	0.495081967
Z Test Statistic	-2.249499716
Lower-Tail Test	
Lower Critical Value	-1.644853
p-Value	0.012240321
Reject the null hypothesis	

9.78 The *American Educational Research Journal* (Fall, 1998) published a study to compare the mathematics achievement test scores of male and female students. The researchers hypothesized that the distribution of test scores for males is more variable than the corresponding distribution for females. Use the summary information in the table to test this claim at $\alpha = .01$.

	Males	Females
Sample size	1,764	1,739
Mean	48.9	48.4
Standard deviation	12.96	11.85

PHStat Output

SBE Exercise 9.78	
Level of Significance	0.01
Population 1 Sample	
Sample Size	1764
Sample Standard Deviation	12.96
Population 2 Sample	
Sample Size	1739
Sample Standard Deviation	11.85
F-Test Statistic	1.196116007
Population 1 Sample Degrees of Freedom	1763
Population 2 Sample Degrees of Freedom	1738
Upper-Tail Test	
Upper Critical Value	1.117699711
p-Value	0.0000912
Reject the null hypothesis	

Chapter 10
Simple Linear Regression

10.1 Introduction

Chapters 10 and 11 in *Statistics for Business and Economics* introduce the topic of regression analysis to the reader. Chapter 10 serves as the introduction of the general concepts of simple linear regression. Simple Linear Regression is how the text introduces the theories and concepts of mathematical modeling to the reader. These topics are then expanded in Chapter 11 of the text.

We will take a similar approach to regression as does the text. We will use Chapter 10 to introduce you to the methods PHStat offers to work with regression analysis. We will see how PHStat can be used to calculate both the correlation and the linear modeling ideas that are presented in the text. We will use the chapter examples that are given in the text to illustrate these methods. The following examples from *Statistics for Business and Economics* are solved with PHStat in this chapter:

Excel Companion		Statistics for Business and Economics	
Example	Page	SBE	Page(s)
10.1	121	Example 10.1	491
10.2	124	Example 10.2	494-495
10.3	126	Example 10.3/10.4	503-506

10.2 The Coefficient of Correlation

Regression analysis is all about the relationship between variables. Chapters 10 and 11 spend time developing the mathematical modeling of one variable using the values of other related variables. The simplest form of this modeling idea is the linear relationship between two variables. This idea, known as correlation, is studied in Chapter 10 of *Statistics for Business and Economics*. We examine how PHStat calculates correlations below.

Example 10.1: Use Example 10.1 found on page 491 of *Statistics for Business and Economics* text.

Legalized gambling is available on several riverboat casinos operated by a city in Mississippi. The mayor of the city wants to know the correlation between the number of casino employees and the yearly crime rate. The records for the past 10 years are examined and the results listed in Table 10.1 are obtained. Calculate the coefficient of correlation r for the data.

Table 10.1

	Number of	Crime
1991	15	1.35
1992	18	1.63
1993	24	2.33
1994	22	2.41
1995	25	2.63
1996	29	2.93
1997	30	3.41
1998	32	3.26
1999	35	3.63
2000	38	4.15

Chapter 10: Simple Linear Regression

Solution:

We begin by **opening** the data file SBE Example 10.1 that is found on the floppy disk included with this manual. Choose the **Regression** option within the **PHStat** menu. Select the **Simple Linear Regression option** from those listed. You should now be looking at the Simple Linear Regression menu shown in Figure 10.1.

To generate the coefficient of correlation, the user needs to enter the location of the two variables of interests (labeled as the Y variable and the X variable), and select the Regression Statistics Table option. Begin by **entering** the column where the dependent variable is located in the **Y Variable Cell Range** of this menu (see Figure 10.2). **Enter** the column where the independent variable is located in the **X Variable Cell Range**. This can be done by **typing** the location or by **clicking and dragging** over the appropriate data cells in your workbook. You have the option of including the variable name in the first cell of data. The user can also specify an optional title for the output. **Check** the **Regression Statistics Table** option and **click OK**. The output generated by PHStat is shown in Table 10.2.

Figure 10.1

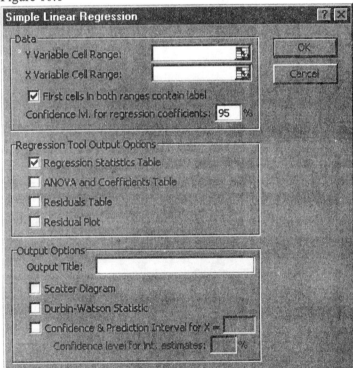

Section 10.3: The Coefficient of Determination and Regression Output

Figure 10.2

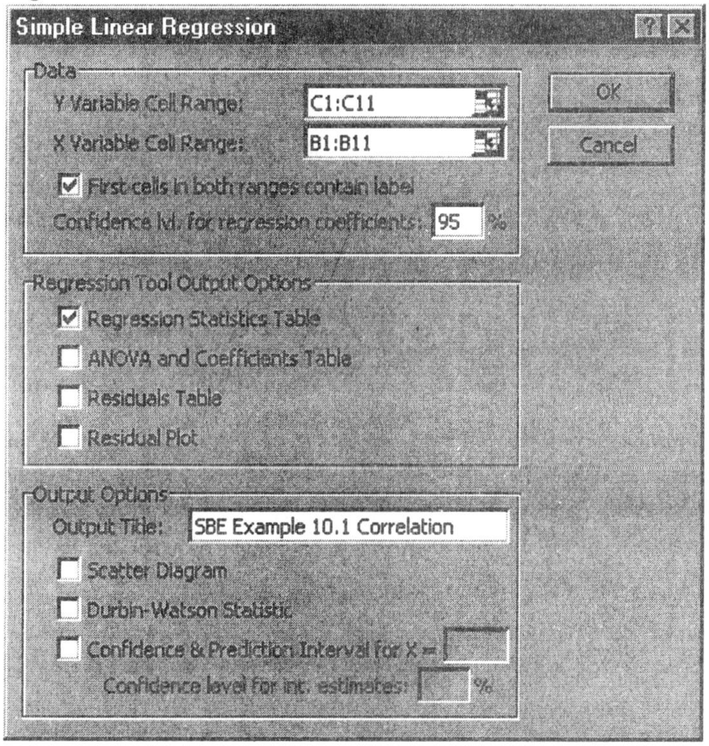

[Simple Linear Regression dialog box screenshot]

Table 10.2

SBE Example 10.1 Correlation	
Regression Statistics	
Multiple R	0.987029777
R Square	0.974227782
Adjusted R Square	0.971006254
Standard Error	0.149932817
Observations	10

From the printout, we find that the coefficient of correlation is r = 0.98703. We next look at how the coefficient of determination and other simple linear regression output are generated within PHStat.

10.3 The Coefficient of Determination and Regression Output

After studying the topic of correlation, the next step in learning regression analysis is understanding the modeling concepts. Our goal in regression is to build a mathematical relationship that attempts to predict the value of one variable, y, with the values of other related variables, the x's. Chapter 10 presents the simplest form of this modeling idea -- simple linear regression. In it, a single independent variable, x, is hypothesized to have a straight-line relationship with the dependent variable, y.

Chapter 10: Simple Linear Regression

The example that we use from *Statistics for Business and Economics* asks the reader to calculate the coefficient of determination from the data. Our purpose here is to use the data from the example to generate the basic simple linear regression output with Excel. As you will see, the coefficient of determination is one of the components of this output.

Example 10.2: We use Example 10.2 found on pages 494-495 of the *Statistics for Business and Economics* text.

Calculate the coefficient of determination for the advertising-sales example that is used as an example throughout the text. The data are shown below in Table 10.3 for convenience.

Table 10.3

Advertising Expenditures, x (100s)	Sales Revenues, y ($1,000s)
1	1
2	1
3	2
4	2
5	4

Solution:

We begin by **opening** the data file SBE Example 10.2 that is found on the floppy disk included with this manual. Choose the **Regression** option within the **PHStat** menu. Select the **Simple Linear Regression option** from those listed. You should now be looking at the same Simple Linear Regression menu shown in Figure 10.1.

To generate the coefficient of determination and other simple regression output, the user needs to enter the location of the two variables of interests (labeled as the Y variable and the X variable), and select both the Regression Statistics Table and the ANOVA and Coefficients Table options. Begin by **entering** the column where the dependent variable is located in the **Y Variable Cell Range** of this menu (see Figure 10.3). **Enter** the column where the independent variable is located in the **X Variable Cell Range**. This can be done by **typing** the location or by **clicking and dragging** over the appropriate data cells in your workbook. You have the option of including the variable name in the first cell of data. The user can also specify an optional title for the output. **Check** both the **Regression Statistics Table** and the **ANOVA and Coefficients Table** options and **click OK**. The output generated by PHStat is shown in Table 10.4.

Section 10.3: The Coefficient of Determination and Regression Output

Figure 10.3

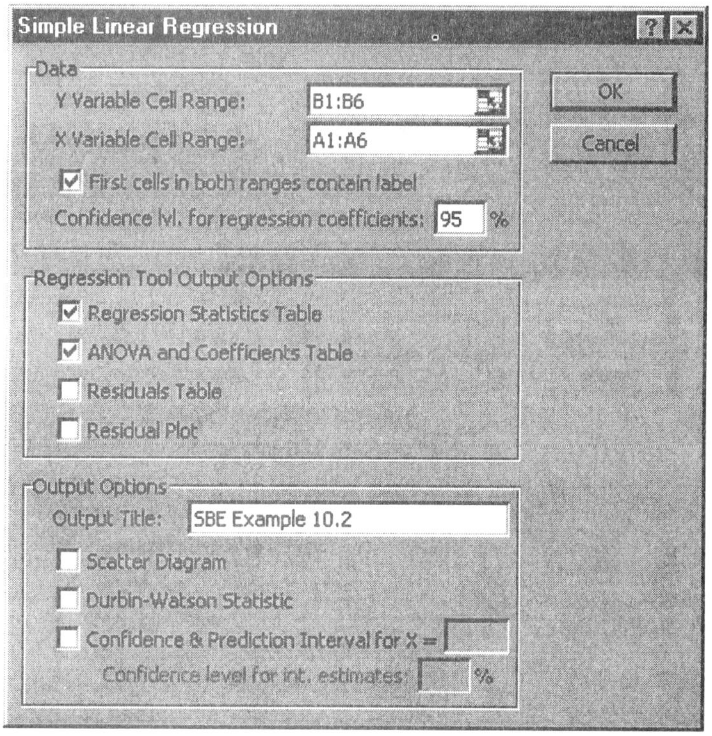

[Simple Linear Regression dialog box showing:
- Data: Y Variable Cell Range: B1:B6, X Variable Cell Range: A1:A6
- First cells in both ranges contain label (checked)
- Confidence lvl. for regression coefficients: 95 %
- Regression Tool Output Options: Regression Statistics Table (checked), ANOVA and Coefficients Table (checked), Residuals Table (unchecked), Residual Plot (unchecked)
- Output Options: Output Title: SBE Example 10.2
- Scatter Diagram (unchecked), Durbin-Watson Statistic (unchecked), Confidence & Prediction Interval for X = (unchecked)]

Table 10.4

SBE Example 10.2

Regression Statistics	
Multiple R	0.903696
R Square	0.816667
Adjusted R Square	0.75556
Standard Error	0.60553
Observations	5

ANOVA

	Df	SS	MS	F	Significance F
Regression	1	4.9	4.9	13.3636	0.035352847
Residual	3	1.1	0.366667		
Total	4	6			

	Coefficients	Standard Error	t Stat	P-value	Lower 95%	Upper 95%
Intercept	-0.1	0.6350852	-0.15746	0.88488	-2.12112675	1.92112675
Advertising Expenditure, x ($100's)	0.7	0.1914854	3.655631	0.03535	0.090607356	1.309392644

Table 10.4 gives the standard simple linear regression output. In addition to the desired coefficient of determination, the simple linear regression model is given. We point out the most important features of the printout generated by PHStat. Compare these values to those shown on the indicated pages of the text.

Excel Printout Values
Multiple R = 0.903696
R Square = 0.816667
Standard Error = 0.605530
Intercept Coefficient = -0.1
X Variable 1 Coefficient = 0.7
T Stat for X Variable 1 = 3.655631
P-value for X Variable 1 = 0.03535

Description of Values
Coefficient of correlation (page 491)
Coefficient of determination (page 495)
Square Root of MSE or s (page 475)
Estimate of β_0 (page 463)
Estimate of β_1 (page 463)
Test Statistic for testing β_1 (page 481)
P-value for testing β_1 (page 481)

We find the value of the coefficient of determination is found on the PHStat to be $R^2 = 0.816667$. Many of the other values presented in Table 10.4 will be discussed in Chapter 11. The last topic that we cover in Chapter 10 is the topic of generating confidence and prediction intervals for specified values of X.

10.4 Estimating and Predicting with a Simple Linear Model

The final step in the simple linear regression analysis is to use the model to estimate and predict values of the dependent variable, y, for specified settings of the independent variable, x. We illustrate this procedure using the following example.

Example 10.3: Use Examples 10.3 and 10.4 found on pages 503 - 506 of the *Statistics for Business and Economics* text.

Example 10.3: Find a 95% confidence interval for the mean monthly sales when the appliance store spends $400 on advertising.

Example 10.4: Predict the monthly sales for the next month, if $400 is spent on advertising. Use a 95% prediction interval.

Solution:

Both of these problems refer back to the data in SBE Example 10.4. We begin by **opening** the data file SBE Example 10.2 that is found on the floppy disk included with this manual. Choose the **Regression** option within the **PHStat** menu. Select the **Simple Linear Regression option** from those listed. You should now be looking at the same Simple Linear Regression menu shown in Figure 10.1.

To generate the confidence and prediction output desired in these two examples, the user needs to enter the location of the two variables of interests (labeled as the Y variable and the X variable), and select only the **Confidence and Prediction Interval for X =** option. Begin by **entering** the column where the dependent variable is located in the **Y Variable Cell Range** of this menu (see Figure 10.4). **Enter** the column where the independent variable is located in the **X Variable Cell Range**. This can be done by **typing** the location or by **clicking and dragging** over the appropriate data cells in your workbook. You have the option of including the variable name in the first cell of data. The user can also specify an optional title for the output. **Check** the **Confidence and Prediction Interval for X =** option. Once this option is selected, the user must then enter a value of X and a confidence level that is desired. In this example, we choose to enter the value of **X = 4** (representing an advertising expenditure of $400) and a **confidence level** of **95%**. We then **click OK**. The output generated by PHStat is shown in Table 10.5.

Section 10.4: Estimating and Predicting with a Simple Linear Model 127

Figure 10.3

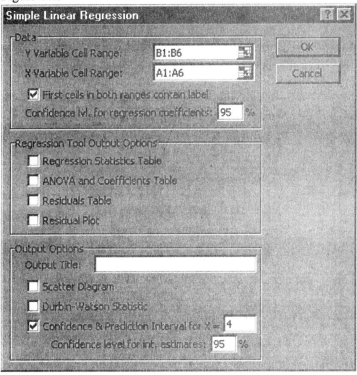

Table 10.5

Confidence Interval Estimate	
X Value	4
Confidence Level	95%
Sample Size	5
Degrees of Freedom	3
t Value	3.182449291
Sample Mean	3
Sum of Squared Difference	10.00
Standard Error of the Estimate	0.605530071
h Statistic	0.3
Average Predicted Y (YHat)	2.7
For Average Predicted Y (YHat)	
Interval Half Width	1.055499021
Confidence Interval Lower Limit	1.644500979
Confidence Interval Upper Limit	3.755499021
For Individual Response Y	
Interval Half Width	2.197196425
Prediction Interval Lower Limit	0.502803575
Prediction Interval Upper Limit	4.897196425

We see that the confidence and prediction interval endpoints are the same values that are calculated in the *Statistics for Business and Economics* text found on pages 503 - 506. We see how easy it is to use PHStat to calculate the regression confidence and prediction intervals described in the text.

Technology Lab

The following exercise from the *Statistics for Business and Economics* text is given for you to practice the procedure covered in the text that is available within PHStat. Included with the exercise is the PHStat output that was generated to solve the problem.

10.70 Emotional exhaustion, or *burnout*, is a significant problem for people with careers in the field of human services. Regression analysis was used to investigate the relationship between burnout and aspects of the human service's professional's job and job-related behavior (*Journal of Applied Behavioral Science*, Vol. 22, 1986). Emotional exhaustion was measured with the Maslach Burnout Inventory, a questionnaire. One of the independent variables considered, called *concentration*, was the proportion of social contacts with individuals who belong to a person's work group. The table below lists the values of the emotional exhaustion index (higher values indicate greater exhaustion) and concentration for a sample of 25 human services professionals who work in a large public hospital.

Exhaustion Index, y	Concentration, x	Exhaustion Index, y	Concentration, x
100	20	493	86
525	60	892	83
300	38	527	79
980	88	600	75
310	79	855	81
900	87	709	75
410	68	791	77
296	12	718	77
120	35	684	77
501	70	141	17
920	80	400	85
810	92	970	96
506	77		

a. Construct a scattergram for the data. Do the variables x and y appear to be related?
b. Find the correlation coefficient for the data and interpret its value. Does your conclusion mean that concentration causes emotional exhaustion? Explain.
c. Test the usefulness of the straight-line relationship with concentration for predicting burnout. Use $\alpha = .05$.
d. Find the coefficient of determination for the model and interpret it.
e. Find a 95% confidence interval for the slope β_1. Interpret the result.
f. Use a 95% confidence interval to estimate the mean exhaustion level for all professionals who have 80% of their social contacts within their work groups. Interpret the interval.

PHStat Output

SBE Exercise 10.70

Regression Statistics

Multiple R	0.78250049
R Square	0.61230702
Adjusted R Square	0.59545081
Standard Error	174.207422
Observations	25

ANOVA

	df	SS	MS	F	Significance F
Regression	1	1102408.2	1102408	36.325295	3.80242E-06
Residual	23	698009.2	30348.2		
Total	24	1800417.4			

	Coefficients	Standard Error	t Stat	P-value	Lower 95%	Upper 95%
Intercept	-29.4967176	106.69716	-0.27645	0.7846695	-250.216316	191.2228804
Concentration, x	8.86547138	1.4709478	6.02705	3.802E-06	5.822588168	11.90835458

Confidence Interval Estimate

X Value	80
Confidence Level	95%
Sample Size	25
Degrees of Freedom	23
t Value	2.068654794
Sample Mean	68.56
Sum of Squared Difference	14026.16
Standard Error of the Estimate	174.207422
h Statistic	0.049330679
Average Predicted Y (YHat)	679.7409925

For Average Predicted Y (YHat)

Interval Half Width	80.04113265
Confidence Interval Lower Limit	599.6998599
Confidence Interval Upper Limit	759.7821252

For Individual Response Y

Interval Half Width	369.1567919
Prediction Interval Lower Limit	310.5842006
Prediction Interval Upper Limit	1048.897784

Chapter 11
Multiple Regression and Model Building

11.1 Introduction

Chapter 11 in *Statistics for Business and Economics* introduces the topic of **multiple** regression analysis to the reader. While Chapter 10 served as the introduction to the general concepts of simple linear regression, Chapter 11 expands these concepts to modeling with several variables. In addition, Chapter 11 examines some of the problems associated with regression analysis and gives methods of detecting and solving these problems.

We utilize Chapter 11 examples to build on the linear regression base developed in the preceding chapter. Through the use of the Regression data analysis tool, PHStat allows the user to build more sophisticated models than the linear models of Chapter 10. We examine both the model building methods and the residual analysis options offered within PHStat. We will use the chapter examples that are given in the text to illustrate these methods. The following examples from *Statistics for Business and Economics* are solved with PHStat in this chapter:

Excel Companion		Statistics for Business and Economics	
Example	Page	SBE	Page(s)
11.1	131	Example 11.3	544-546
11.2	135	Example 11.4	559
11.3	137	Example 11.5	577-578
11.4	140	Example 11.11	616-618
11.5	143	Example 11.14/11.15	640-643

11.2 Multiple Regression Model Building

We have seen in Chapter 10 how to use PHStat to build a simple linear regression model using one independent variable, x. The next step in our regression process in to add more independent variables into the regression model. PHStat allows for this using the same menus as seen in the simple linear regression chapter. We us an example from the text below.

Example 11.1: We use Example 11.3 found on pages 544-546 in the *Statistics for Business and Economics* text.

A collector of antique grandfather clocks knows that the price received for the clocks increases linearly with the age of the clocks. Moreover, the collector hypothesizes that the auction price of the clocks will increase linearly as the number of bidders increase. Thus, the following model is hypothesized:

$$y = \beta_0 + \beta_1 x_1 + \beta_2 x_2 + \varepsilon$$

Chapter 11: Multiple Regression and Model Building

A sample of 32 auction prices of grandfather clocks, along with their age and the number of bidders is shown in Table 11.1 The model $y = \beta_0 + \beta_1 x_1 + \beta_2 x_2 + \varepsilon$ is fit to the data. Use PHStat to:

a. Test the hypothesis that the mean auction price of a clock increases as the number of bidders increases when age is held constant, that is, $\beta_2 > 0$. Use $\alpha = .05$.
b. Interpret the estimates of the β coefficients in the model.

Table 11.1

Age(x_1)	Number of Bidders (x_2)	Auction Price (y)	Age(x_1)	Number of Bidders (x_2)	Auction Price (y)
127	13	1,235	170	14	2,131
115	12	1,080	182	8	1,550
127	7	845	162	11	1,884
150	9	1,522	184	10	2,041
156	6	1,047	143	6	845
182	11	1,979	159	9	1,483
156	12	1,822	108	14	1,055
132	10	1,253	175	8	1,545
137	9	1,297	108	6	729
113	9	946	179	9	1,792
137	15	1,713	111	15	1,175
117	11	1,024	187	8	1,593
137	8	1,147	111	7	785
153	6	1,092	115	7	744
117	13	1,152	194	5	1,356
126	10	1,336	168	7	1,262

Solution:

We need to generate the multiple regression model hypothesized above using PHStat. The printout generated must include the individual coefficient estimates and the corresponding t-tests of those parameters. Fortunately, the standard PHStat regression output yields both of the desired values.

We begin by **opening** the data file SBE Example 11.3 that is found on the floppy disk included with this manual. Choose the **Regression** option within the **PHStat** menu. Select the **Multiple Regression** option from those listed. You should now be looking at the same Multiple Regression Menu shown in Figure 11.1.

Figure 11.1

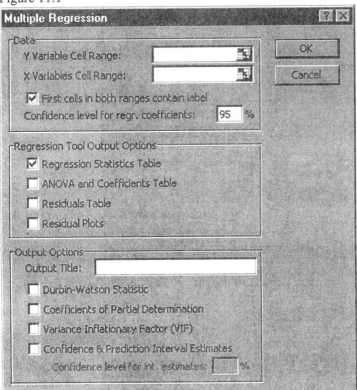

To generate the coefficient estimate information that is desired in this problem, the user needs to enter the location of the two variables of interests (labeled as the Y variable and the X variable), and select the ANOVA and Coefficients Table option. Begin by **entering** the column where the dependent variable is located in the **Y Variable Cell Range** of this menu (see Figure 11.2). **Enter** the column where the independent variables are located in the **X Variables Cell Range**. This can be done by **typing** the location or by **clicking and dragging** over the appropriate data cells in your workbook. You have the option of including the variable name in the first cell of data. The user can also specify an optional title for the output. **Check** the **ANOVA and Coefficients Table** option and **click OK**. The output generated by PHStat is shown in Table 11.2.

Figure 11.2

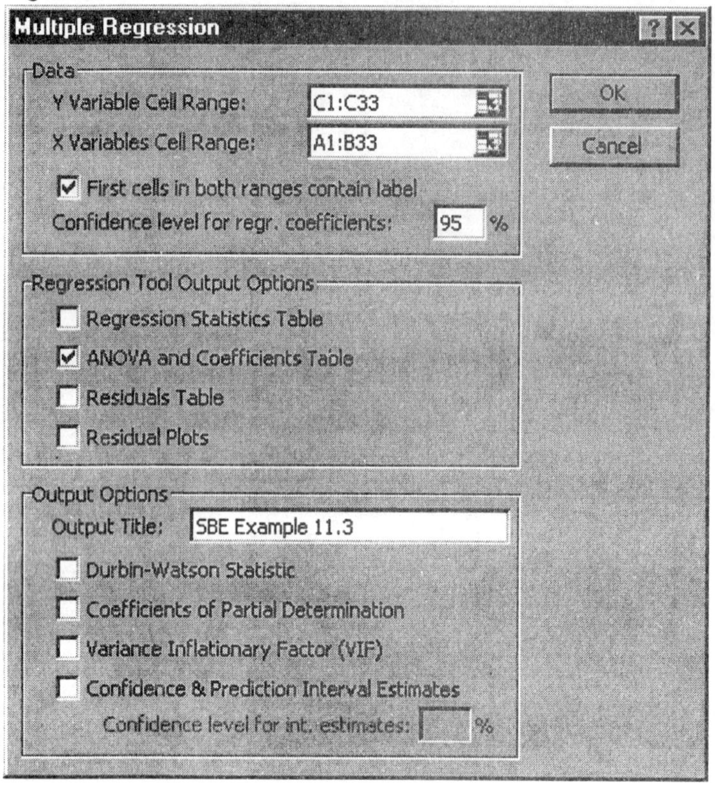

[Multiple Regression dialog box]
- Y Variable Cell Range: C1:C33
- X Variables Cell Range: A1:B33
- ☑ First cells in both ranges contain label
- Confidence level for regr. coefficients: 95 %

Regression Tool Output Options:
- ☐ Regression Statistics Table
- ☑ ANOVA and Coefficients Table
- ☐ Residuals Table
- ☐ Residual Plots

Output Options:
- Output Title: SBE Example 11.3
- ☐ Durbin-Watson Statistic
- ☐ Coefficients of Partial Determination
- ☐ Variance Inflationary Factor (VIF)
- ☐ Confidence & Prediction Interval Estimates
- Confidence level for int. estimates: ___ %

Table 11.2

SBE Example 11.3

ANOVA

	df	SS	MS	F	Significance F
Regression	2	4283062.96	2141531	120.1882	9.2164E-15
Residual	29	516726.5399	17818.16		
Total	31	4799789.5			

	Coefficients	Standard Error	t Stat	P-value	Lower 95%	Upper 95%
Intercept	-1338.95	173.8094707	-7.70356	1.71E-08	-1694.4318	-983.470865
Age (x1)	12.74057	0.904740307	14.08202	1.69E-14	10.8901714	14.5909768
Number of Bidders (x2)	85.95298	8.728523289	9.847368	9.34E-11	68.1011401	103.8048287

Compare this output to the MINITAB output shown on page 545 of the text. In order to test whether the price of a clock increases as the number of bidders increases, holding age constant, we use the test statistic and p-value shown on the printout for the x_2 variable. The printout shows that $t = 9.847$ and $p \approx 0$. The estimates of the β coefficients can be found in the Coefficients column in the printout. Our estimates of β_0, β_1, and β_2 are -1338.95, 12.74 and 85.95, respectively. We refer you to the text for more detailed information regarding the interpretations and conclusion that should be made for these values.

Section 11.2: Multiple Regression Model Building

The next step of a regression analysis is to test all the hypothesized variables simultaneously. We refer to this process as checking the usefulness of the model. This process is illustrated in the following example.

Example 11.2: We use Example 11.4 found on page 559 in the *Statistics for Business and Economics* text.

A collector of antique grandfather clocks knows that the price received for the clocks increases linearly with the age of the clocks. Moreover, the collector hypothesizes that the auction price of the clocks will increase linearly as the number of bidders increase. Thus, the following model is hypothesized:

$$y = \beta_0 + \beta_1 x_1 + \beta_2 x_2 + \varepsilon$$

A sample of 32 auction prices of grandfather clocks, along with their age and the number of bidders is shown in Table 11.1 The model $y = \beta_0 + \beta_1 x_1 + \beta_2 x_2 + \varepsilon$ is fit to the data. Use Excel to :

a. Find and interpret the adjusted coefficient of determination R_a^2 for this example.
b. Conduct the global F-test of model usefulness at the $\alpha = .05$ level of significance.

Solution:

We need to generate the multiple regression model hypothesized above using PHStat. The printout generated must include the adjusted coefficient of determination and the global-F test and p-value information. Fortunately, the standard PHStat regression output yields both of the desired values.

We begin by **opening** the data file SBE Example 11.3 that is found on the floppy disk included with this manual. Choose the **Regression** option within the **PHStat** menu. Select the **Multiple Regression** option from those listed. You should now be looking at the same Multiple Regression Menu shown in Figure 11.1.

To generate the adjusted coefficient of determination and the global-F test information desired in this problem, the user needs to enter the location of the two variables of interests (labeled as the Y variable and the X variable), and select both the Regression Statistics Table and the ANOVA and Coefficients Table options. Begin by **entering** the column where the dependent variable is located in the **Y Variable Cell Range** of this menu (see Figure 11.3). **Enter** the column where the independent variables are located in the **X Variables Cell Range**. This can be done by **typing** the location or by **clicking and dragging** over the appropriate data cells in your workbook. You have the option of including the variable name in the first cell of data. The user can also specify an optional title for the output. **Check** both the **Regression Statistics Table** and the **ANOVA and Coefficients Table** options and **click OK**. The output generated by PHStat is shown in Table 11.3.

136   Chapter 11:   Multiple Regression and Model Building

Figure 11.3

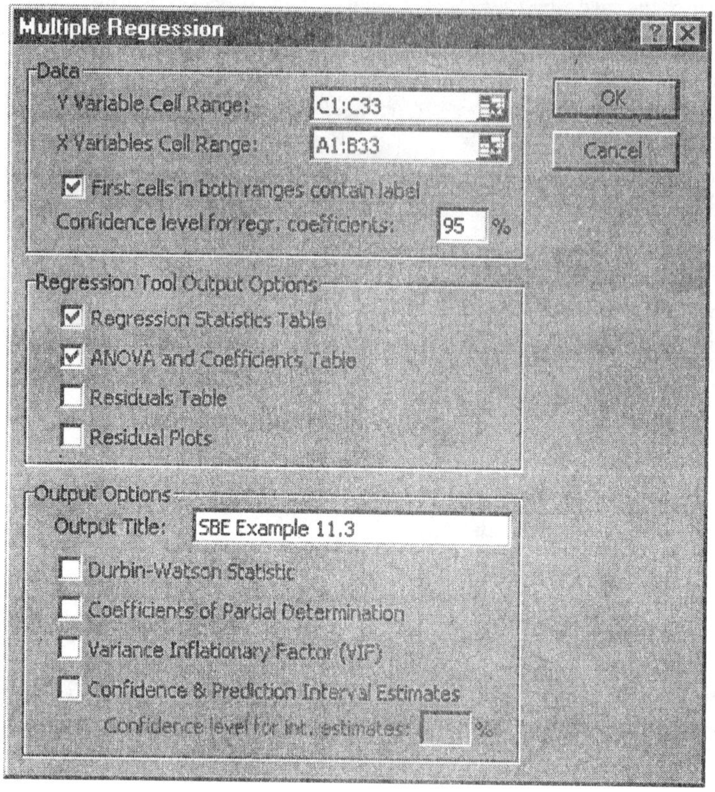

[Screenshot of Multiple Regression dialog box]

Data
- Y Variable Cell Range: C1:C33
- X Variables Cell Range: A1:B33
- ☑ First cells in both ranges contain label
- Confidence level for regr. coefficients: 95 %

Regression Tool Output Options
- ☑ Regression Statistics Table
- ☑ ANOVA and Coefficients Table
- ☐ Residuals Table
- ☐ Residual Plots

Output Options
- Output Title: SBE Example 11.3
- ☐ Durbin-Watson Statistic
- ☐ Coefficients of Partial Determination
- ☐ Variance Inflationary Factor (VIF)
- ☐ Confidence & Prediction Interval Estimates
  - Confidence level for int. estimates: __ %

Table 11.3

SBE Example 11.3						
Regression Statistics						
Multiple R	0.94463957					
R Square	0.892343916					
Adjusted R Square	0.884919359					
Standard Error	133.4846678					
Observations	32					

ANOVA

	df	SS	MS	F	Significance F	
Regression	2	4283062.96	2141531	120.1882	9.2164E-15	
Residual	29	516726.5399	17818.16			
Total	31	4799789.5				

	Coefficients	Standard Error	t Stat	P-value	Lower 95%	Upper 95%
Intercept	-1338.95134	173.8094707	-7.70356	1.71E-08	-1694.4318	-983.47087
Age (x1)	12.7405741	0.904740307	14.08202	1.69E-14	10.8901714	14.590977
Number of Bidders (x2)	85.95298437	8.728523289	9.847368	9.34E-11	68.1011401	103.80483

Section 11.2: Multiple Regression Model Building 137

The adjusted coefficient of determination R_a^2 is listed as the Adjusted R Square value in the Regression Statistics table above. The R_a^2 value of $R_a^2 = 0.8849$ can be compared to the value shown on page 559 of the text. The global F statistic and p-value are both shown in the ANOVA table in the printout above. The global F statistic of $F = 120.188$ and the p-value of $p \approx 0$ are identical to the values shown in the text on page 559. We refer you to the text for more detailed information regarding the interpretations and conclusions that should be made for these values.

The next step in the model building process of a regression analysis is to add various types of regression terms to the model. Whether the terms added are interactions, quadratics, or qualitative terms, the process within PHStat is the same. We illustrate this process by adding an interaction component to the preceding example to illustrate. Please note that the process of adding quadratic and qualitative terms to the regression model is identical to the process demonstrated in the next example.

Example 11.3: We use Example 11.5 found on pages 577-578 of the *Statistics for Business and Economics* text.

Refer to Examples 11.3 and 11.4. Suppose the collector of grandfather clocks, having observed many auctions, believes that the *rate of increase* of the auction price with age will be driven upward by a large number of bidders. Thus, instead of a relationship in which the rate of the price is the same for any number of bidders, the collector believes the slope of the price-age relationship increases as the number of bidders increases. Consequently, the interaction model is proposed:

$$Y = \beta_0 + \beta_1 x_1 + \beta_2 x_2 + \beta_3 x_1 x_2 + \varepsilon$$

a. Test the overall utility of the model using the global F-test at $\alpha = .05$.
b. Test the hypothesis (at $\alpha = .05$) that the price-age slope increases as the number of bidders increases - that is, that age and number of bidders, x_2, interact positively.

Solution:

In order to fit a regression model that includes an interaction term, this term must be added to the data set in our Excel file. We refer the reader to Sections P.9 and P.10 to use formulas within Excel to create the desired age*bidder interaction term. Once created, the data look like the data shown in Table 11.4 below:

Chapter 11: Multiple Regression and Model Building

Table 11.4

Age (x1)	Number of Bidders (x2)	Age*Bidders (x1*x2)	Auction Price (y)
127	13	1651	$1,235
115	12	1380	$1,080
127	7	889	$845
150	9	1350	$1,522
156	6	936	$1,047
182	11	2002	$1,979
156	12	1872	$1,822
132	10	1320	$1,253
137	9	1233	$1,297
113	9	1017	$946
137	15	2055	$1,713
117	11	1287	$1,024
137	8	1096	$1,147
.	.	.	.
.	.	.	.
.	.	.	.

Once the data set includes the interaction variable, choose the **Regression** option within the **PHStat** menu. Select the **Multiple Regression** option from those listed. You should now be looking at the same Multiple Regression Menu shown in Figure 11.1.

To generate the coefficient estimate and the global-F test information desired in this problem, the user needs to enter the location of the two variables of interests (labeled as the Y variable and the X variable), and select both the Regression Statistics Table and the ANOVA and Coefficients Table options. Begin by **entering** the column where the dependent variable is located in the **Y Variable Cell Range** of this menu (see Figure 11.4). **Enter** the column where the independent variables are located in the **X Variables Cell Range**. This can be done by **typing** the location or by **clicking and dragging** over the appropriate data cells in your workbook. You have the option of including the variable name in the first cell of data. The user can also specify an optional title for the output. **Check** both the **Regression Statistics Table** and the **ANOVA and Coefficients Table** options and **click OK**. The output generated by PHStat is shown in Table 11.5.

Section 11.2: Multiple Regression Model Building

Figure 11.4

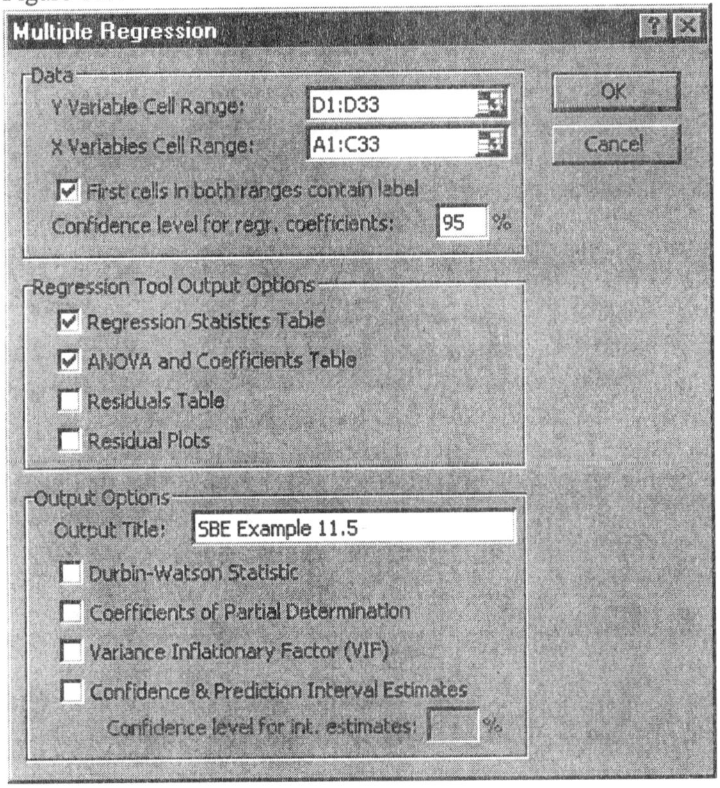

Table 11.5

SBE Example 11.5

Regression Statistics	
Multiple R	0.976668248
R Square	0.953880866
Adjusted R Square	0.948939531
Standard Error	88.91451215
Observations	32

ANOVA

	df	SS	MS	F	Significance F
Regression	3	4578427.37	1526142.46	193.0411	8.35E-19
Residual	28	221362.133	7905.79047		
Total	31	4799789.5			

	Coefficients	Standard Error	t Stat	P-value	Lower 95%	Upper 95%
Intercept	320.4579934	295.141285	1.08577827	0.286837	-284.1122	925.02819
Age (x1)	0.878142475	2.03215593	0.43212357	0.6689613	-3.2845449	5.0408299
Number of Bidders (x2)	-93.26482436	29.8916162	-3.1200998	0.0041646	-154.49509	-32.03456
Age*Bidders (x1*x2)	1.297845824	0.2123326	6.11232488	1.353E-06	0.8629017	1.7327899

140 Chapter 11: Multiple Regression and Model Building

11.3 Comparing Two Regression Models

We have seen how PHStat can be used to fit regression models with just quantitative variables and regression models with just qualitative variables. For more complicated models, PHStat allows the user to input both quantitative and qualitative variables into a single multiple regression model. The variables must be located in adjacent columns of data within the PHStat worksheet. By specifying the appropriate columns in the Regression data analysis menu, any number of quantitative and qualitative variables can be combined.

The final step in the model building topic is to develop a method that allows the user to compare two regression models to determine which is the better predictor of the dependent variable. Section 11.11 in the text details the partial-F test for testing a portion of the regression model. By fitting two separate models within PHStat, it is possible to calculate the partial-F test statistic that the book details. We demonstrate with the following example.

Example 11.4: We use Example 11.11 found on pages 616 - 617 of the *Statistics for Business and Economics* text.

Many companies manufacture products (e.g., steel, paint, and gasoline) that are at least partially chemically produced. In many instances, the quality of the finished product is a function of the temperature and pressure at which the chemical reactions take place. Suppose you want to model the quality y of a product as a function of the temperature x_1 and the pressure x_2 at which it is produced. Four inspectors independently assign a quality score between 0 and 100 to each product, and then the quality y is calculated by averaging the four scores. An experiment is conducted by varying temperature between 80°F and 100°F and pressure between 50 and 60 pounds per square inch (psi). The resulting data are given below.

x_1 (°F)	x_2 (psi)	y	x_1 (°F)	x_2 (psi)	y	x_1 (°F)	x_2 (psi)	y
80	50	50.8	90	50	63.4	100	50	46.6
80	50	50.7	90	50	61.6	100	50	49.1
80	50	49.4	90	50	63.4	100	50	46.4
80	55	93.7	90	55	93.8	100	55	69.8
80	55	90.9	90	55	92.1	100	55	72.5
80	55	90.9	90	55	97.4	100	55	73.2
80	60	74.5	90	60	70.9	100	60	38.7
80	60	73.0	90	60	68.8	100	60	42.5
80	60	71.2	90	60	71.3	100	60	41.4

a. Fit a complete second-order model to the data.
b. Sketch the fitted model in three dimensions.
c. Do the data provide sufficient evidence to indicate that the second-order terms, β_3, β_4, and β_5 contribute information for the prediction of y?

Solution:

In order to compare the determine if the second-order terms contribute information for predicting y, both a full model (containing the second-order terms) and a reduced model (that does not contain the second-order terms) must be fit in PHStat. We utilize the procedures covered in the last section to fit both models (using Data File SBE Table 11.7). The corresponding regression output for both models is shown below in Table 11.6 and Table 11.7.

Section 11.4: Stepwise Regression

Table 11.6

Regression Analysis - Complete Model					
ANOVA					
	df	SS	MS	F	Significance F
Regression	5	8402.26454	1680.4529	596.32392	7.0235E-22
Residual	21	59.1784259	2.8180203		
Total	26	8461.44296			

	Coefficients	Standard Error	t Stat	P-value	Lower 95%	Upper 95%
Intercept	-5127.8991	110.296015	-46.492152	1.153E-22	-5357.2722	-4898.53
x1 (oF)	31.0963889	1.34441322	23.130083	2.01E-16	28.3005281	33.89225
x2 (psi)	139.747222	3.14005412	44.504718	2.86E-22	133.217121	146.2773
x1*x1	-0.1333889	0.00685325	-19.463602	6.455E-15	-0.147641	-0.11914
x2*x2	-1.1442222	0.02741299	-41.740145	1.084E-21	-1.2012307	-1.08721
x1*x2	-0.1455	0.00969196	-15.01245	1.059E-12	-0.1656555	-0.12534

Table 11.7

Regression Analysis - Reduced Model					
ANOVA					
	df	SS	MS	F	Significance F
Regression	2	1789.934444	894.96722	3.219544	0.0577239
Residual	24	6671.508519	277.97952		
Total	26	8461.442963			

	Coefficients	Standard Error	t Stat	P-value	Lower 95%	Upper 95%
Intercept	106.08519	55.94500427	1.8962405	0.07003	-9.3796049	221.55
x1 (oF)	-0.9161111	0.392979729	-2.331192	0.028469	-1.7271812	-0.105041
x2 (psi)	0.7877778	0.785959458	1.0023135	0.326191	-0.8343625	2.409918

Compare these two printouts versus the SAS printouts fund on pages 617 and 618 in the text. We refer you to the text for more detailed information regarding the interpretations and conclusion that should be made for these values.

11.4 Stepwise Regression

Optional Section 11.12 in the text offers the reader information concerning the topic of stepwise regression. This procedure allows the user to identify, from a large pool of potential independent variables, the set of the most useful variables for predicting a dependent variable To access the Stepwise Regression menu, click on the PHStat, choose the **Regression** option within the **PHStat** menu. Select the **Stepwise Regression** option from those listed. You should now be looking at the same Stepwise Regression menu shown in Figure 11.5.

Figure 11.5

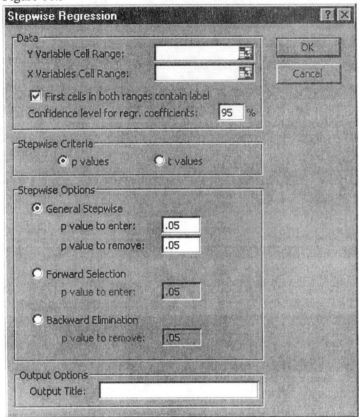

As with all the regression menus, the location of the dependent variable is entered in **the Y Variable Cell Range** and the location of all potential independent variables are entered in the **X Variable Cell Range** in the menu. The user has the option of specifying both a **Stepwise Criteria** and **Stepwise Option** as part of the stepwise process. These options go beyond the scope of the material presented in the text and we defer to some of the references listed in Chapter 11 of the text. For the procedure discussed in the text, the user should specify both a **p values** Stepwise Criteria and a **Forward** Selection Stepwise Option. The user then would **click** on **OK** to finish.

11.5 Residual Analysis

So far we have covered model building, model testing, and stepwise regression within the PHStat program. The last topic to address is the topic of residual analysis. As specified in the text, the topic of residual analysis requires the construction of several different graphical displays that are readily available from the multiple regression menu. The Multiple Regression menu is again shown below in Figure 11.6. Before demonstrating the various plots available, we list the options available and give a brief description regarding their use in a residual analysis of a regression model.

Section 11.5: Residual Analysis

Figure 11.6

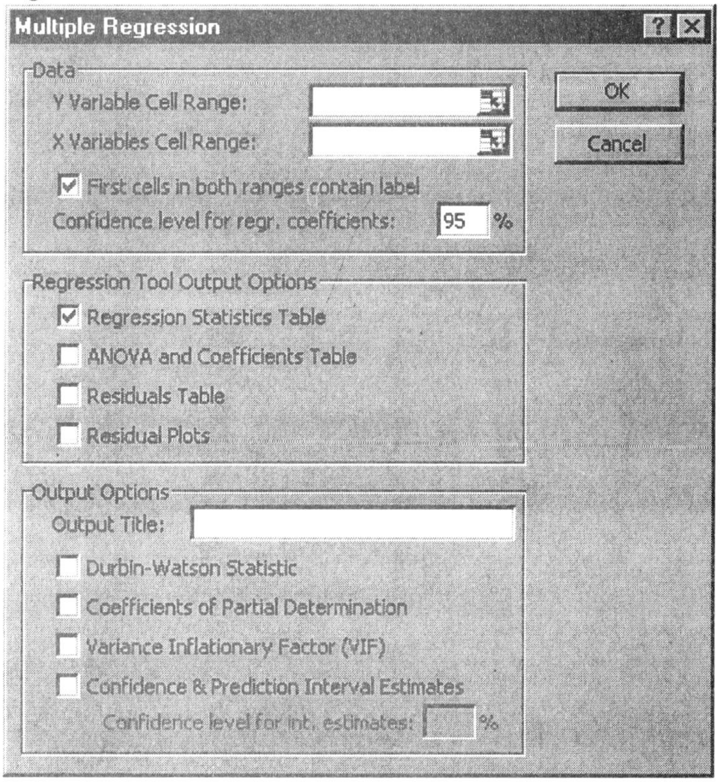

[Multiple Regression dialog box screenshot]

Residual Table Provides a listing of the residuals for the hypothesized regression model

Residual Plots Generates a scatter plot for the regression residuals plotted against each of the Independent variables entered into the regression model.

We illustrate how to use these plots using the following example from the text. In addition, we explain how to generate other graphical displays that are of use in a residual analysis

Example 11.5: We use 11.14 and 11.15 found on pages 640-643 of the *Statistics for Business and Economics* text.

The data for the grandfather clock example used throughout this chapter are repeated in Table 11.8, with one important difference: The auction price of the clock at the top of the second column has been changed from $2,131 to 1,131. The interaction model

$$y = \beta_0 + \beta_1 x_1 + \beta_2 x_2 + \beta_3 x_1 x_2 + \varepsilon$$

is again fit to these (modified) data. Use PHStat to generate all corresponding residual analysis printouts.

144 Chapter 11: Multiple Regression and Model Building

Table 11.7

Age(x_1)	Number of Bidders (x_2)	Auction Price (y)	Age(x_1)	Number of Bidders (x_2)	Auction Price (y)
127	13	1,235	170	14	1,131
115	12	1,080	182	8	1,550
127	7	845	162	11	1,884
150	9	1,522	184	10	2,041
156	6	1,047	143	6	845
182	11	1,979	159	9	1,483
156	12	1,822	108	14	1,055
132	10	1,253	175	8	1,545
137	9	1,297	108	6	729
113	9	946	179	9	1,792
137	15	1,713	111	15	1,175
117	11	1,024	187	8	1,593
137	8	1,147	111	7	785
153	6	1,092	115	7	744
117	13	1,152	194	5	1,356
126	10	1,336	168	7	1,262

Solution: The first step in this solution is to fit the proposed regression model to the modified grandfather clock data. We will follow closely the Excel instructions used in Example 11.3 and change only the requested residual analysis output.

In order to fit a regression model that includes an interaction term, this term must be added to the data set in our Excel file. We refer the reader to Sections P.9 and P.10 to use formulas within Excel to create the desired age*bidder interaction term. Once created, the data look like the data shown in Table 11.8 below:

Table 11.8

Age (x1)	Number of Bidders (x2)	Age*Bidders (x1*x2)	Auction Price (y)
127	13	1651	$1,235
115	12	1380	$1,080
127	7	889	$845
150	9	1350	$1,522
156	6	936	$1,047
182	11	2002	$1,979
156	12	1872	$1,822
132	10	1320	$1,253
137	9	1233	$1,297
113	9	1017	$946
137	15	2055	$1,713
117	11	1287	$1,024
137	8	1096	$1,147
.	.	.	.
.	.	.	.
.	.	.	.

Once the data set includes the interaction variable, choose the **Regression** option within the **PHStat** menu. Select the **Multiple Regression** option from those listed. You should now be looking at the same Simple Linear Regression menu shown in Figure 11.7.

Section 11.5: Residual Analysis

Figure 11.7

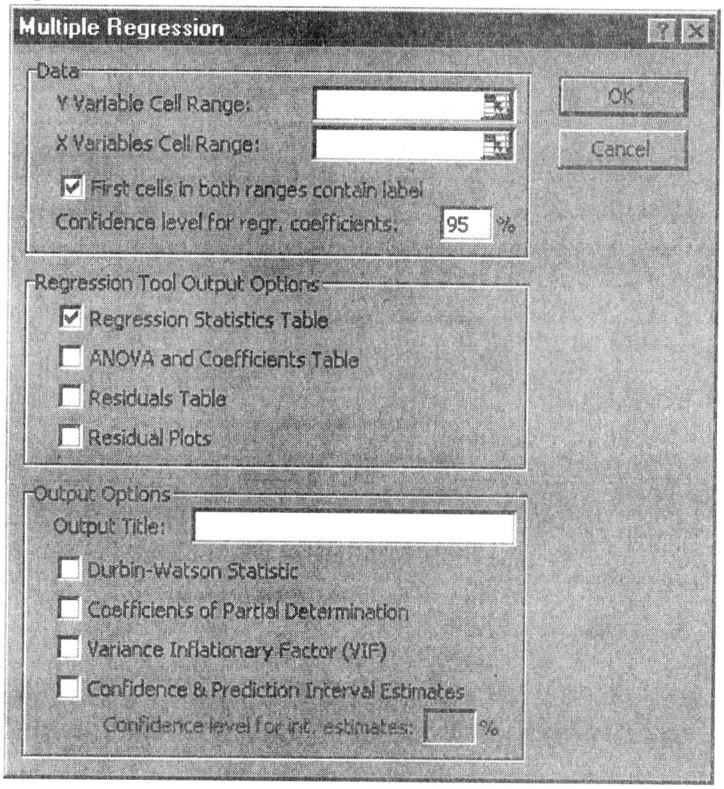

[Multiple Regression dialog box showing Data section with Y Variable Cell Range, X Variables Cell Range, First cells in both ranges contain label (checked), Confidence level for regr. coefficients: 95%. Regression Tool Output Options: Regression Statistics Table (checked), ANOVA and Coefficients Table, Residuals Table, Residual Plots. Output Options: Output Title, Durbin-Watson Statistic, Coefficients of Partial Determination, Variance Inflationary Factor (VIF), Confidence & Prediction Interval Estimates, Confidence level for int. estimates.]

We use this menu to generate both the regression output that we need and the residual analysis information that is desired in this. The user needs to enter the location of the two variables of interests (labeled as the Y variable and the X variable), and select all of the Regression Tool Output Options listed in the menu. Begin by **entering** the column where the dependent variable is located in the **Y Variable Cell Range** of this menu (see Figure 11.8). **Enter** the column where the independent variables are located in the **X Variables Cell Range**. This can be done by **typing** the location or by **clicking and dragging** over the appropriate data cells in your workbook. You have the option of including the variable name in the first cell of data. The user can also specify an optional title for the output. **Check the Regression Statistics Table**, the **ANOVA and Coefficients Table**, the **Residuals Table**, and the **Residual Plots** options and **click OK**. The output generated by PHStat is shown in Tables 11.9 and 11.10 and Figures 11.9, 11.10, and 11.11.

Figure 11.8

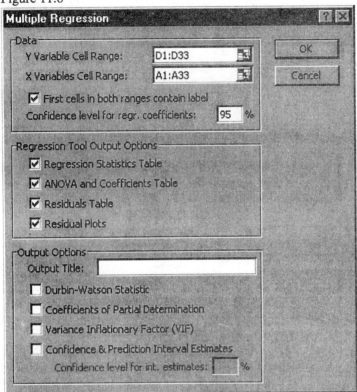

Our purpose in this section is to illustrate the different types of PHStat output that are available and to give the commands necessary to generate these output. We will refer you to section 11.13 of the text for information regarding the assessment and interpretation of these printouts. We point out here the various output generated by the different options within PHStat.

We begin with the standard regression output generated by PHStat (see Table 11.9). This output includes the summary output table, the ANOVA table for the proposed regression model, and the individual coefficient estimates along with their corresponding confidence and test of hypothesis information. This output is provided whenever a regression model is specified in the Regression menu of PHStat. No residual option needs to be specified in order to generate this standard regression output.

Section 11.5: Residual Analysis

Table 11.9

SUMMARY OUTPUT									
Regression Statistics									
Multiple R	0.85391853								
R Square	0.72917685								
Adjusted R Square	0.70016009								
Standard Error	200.597562								
Observations	32								

ANOVA									
	df	SS	MS	F	Sign. F				
Regression	3	3033586.804	1011196	25.1295	4.2774E-08				
Residual	28	1126702.696	40239.38						
Total	31	4160289.5							

	Coefficients	Standard Error	t Stat	P-value	Lower 95%	Upper 95%	Lower 95.0%	Upper 95.0%
Intercept	-512.81017	665.8600588	-0.77015	0.447662	-1876.7642	851.143862	-1876.7642	851.1438618
AGE X1	8.16507852	4.584690573	1.780944	0.085775	-1.2262449	17.556402	-1.22624494	17.55640198
BIDDERS X2	19.8876621	67.4376453	0.294904	0.770242	-118.25225	158.027572	-118.252247	158.0275715
AGE*BID, X1*X2	0.31964386	0.47903768	0.667262	0.510067	-0.6616214	1.30090916	-0.66162145	1.300909165

The Residuals option asks PHStat to compute these values and display them in tabular form. Table 11.10 provides the requested information for this example. We refer the reader to Section 11.13 of *Statistics for Business and Economics* for more details regarding how this table can be used in the residual analysis portion of a regression analysis.

Table 11.10

RESIDUAL OUTPUT		
Observation	Predicted Auction Price (y)	Residuals
1	1310.426418	-75.42641826
2	1105.934329	-25.93432854
3	947.531826	-102.531826
4	1322.459774	199.5402255
5	1179.454702	-132.4547022
6	1831.925406	147.0745936
7	1597.967326	224.0326744
8	1185.786708	67.2132924
9	1178.915422	118.0845776
10	913.9104648	32.08953518
11	1560.988646	152.0113539
12	1072.649945	-48.64994469
13	1115.236552	31.76344818
14	1149.205877	-57.20587718
15	1187.221932	-35.22193162
16	1117.617605	218.382395
17	1914.432829	-783.4328287
18	1597.736874	-47.73687377
19	1598.302187	285.6978126
20	1776.585596	264.4144036
21	1048.376461	-203.3764606
22	1421.836634	61.16336641
23	1130.747092	-75.74709239
24	1522.681268	22.31873186
25	695.4735024	33.52649759
26	1642.674098	149.3259017
27	1224.0355	-49.03550027
28	1651.348021	-58.34802065
29	781.0904577	3.909542288
30	822.7007998	-78.70079979
31	1480.707915	-124.7079148
32	1374.037832	-112.0378323

The Residual Plots option in the Regression menu of Excel plots the residuals calculated by Excel versus the independent variables that were hypothesized in the regression model. Figure 11.9 shows the residuals plotted against independent variables x_1 and x_2. We note here that a residual plot versus the interaction term $x_1 x_2$ is also generated by Excel but not reproduced here. Please refer to Section 11.13 of *Statistics for Business and Economics* for more details regarding how these plots can be used in the residual analysis portion of a regression analysis.

These are the two residual analysis outputs that PHStat generates for the user. There are, however, two additional graphs that are useful to the user in a residual analysis. The first is a residual plot that is helpful for analyzing the equal variance assumption in the regression analysis. The second is a stem-and-leaf display of the residuals that is helpful for checking the normality assumption of the random errors. While neither plot is given automatically by PHStat, both are easy to generate given the tools available within PHStat. We refer the reader to the sections on generating scatterplots and stem-and-leaf displays to find the directions for generating these graphs. These two outputs are shown below in Figures 11.10 and 11.11.

Figure 11.10

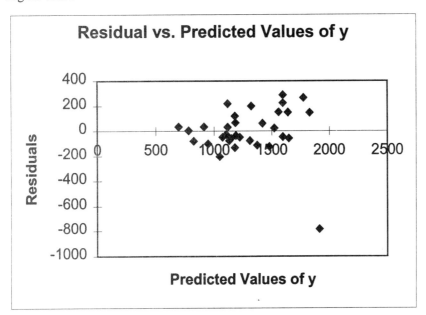

Figure 11.11

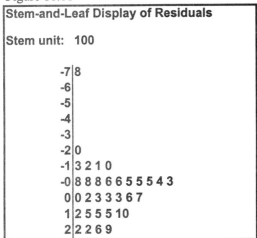

We refer the reader to Section 11.13 of *Statistics for Business and Economics* for more details regarding how these output can be used in the residual analysis portion of a regression analysis.

Chapter 11: Multiple Regression and Model Building

Technology Lab

The following exercise from the *Statistics for Business and Economics* text is given for you to practice the procedure covered in the text that is available within PHStat. Included with the exercise is the PHStat output that was generated to solve the problem.

11.136 A firm that has developed a new type of light bulb is interested in evaluating its performance in order to decide whether to market it. It is known that the light output of the bulb depends on the cleanliness of its surface area and the length of time the bulb has been in operation. Use the data in the table at right and the procedures you learned in this chapter to build a regression model that relates drop in light output to bulb surface cleanliness and length of operation. Be sure to conduct a residual analysis also.

PHStat Output

SBE Exercise 11.136

Regression Statistics

Multiple R	0.92471733
R Square	0.85510214
Adjusted R Square	0.828757074
Standard Error	5.391039945
Observations	14

ANOVA

	df	SS	MS	F	Significance F
Regression	2	1886.660714	943.330357	32.457772	2.4313E-05
Residual	11	319.6964286	29.0633117		
Total	13	2206.357143			

	Coefficients	Standard Error	t Stat	P-value	Lower 95%	Upper 95%
Intercept	12.80357143	2.970318341	4.31050479	0.0012341	6.26594153	19.3412013
Bulb Surface	-17.28571429	2.881632062	-5.9985848	8.944E-05	-23.628147	-10.943282
Length of Operation	0.0096875	0.00180102	5.37889629	0.0002237	0.00572348	0.01365152

RESIDUAL OUTPUT

Observation	Predicted Drop in Output	Residuals
1	12.80357143	-12.8035714
2	16.67857143	-0.67857143
3	20.55357143	1.44642857
4	24.42857143	2.57142857
5	28.30357143	3.69642857
6	32.17857143	3.82142857
7	36.05357143	1.94642857
8	-4.482142857	4.48214286
9	-0.607142857	4.60714286
10	3.267857143	2.73214286
11	7.142857143	0.85714286
12	11.01785714	-2.01785714
13	14.89285714	-3.89285714
14	18.76785714	-6.76785714

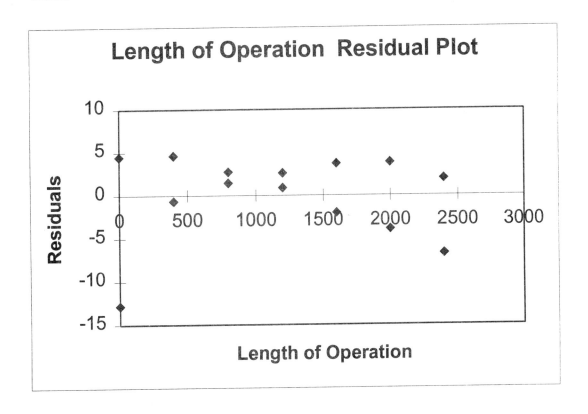

152 Chapter 11: Multiple Regression and Model Building

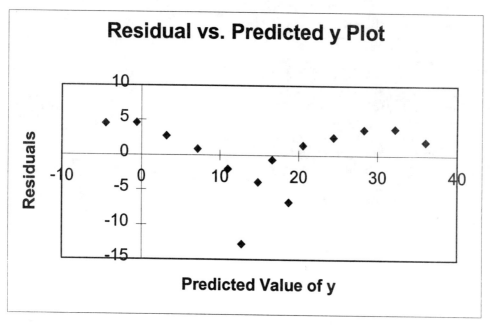

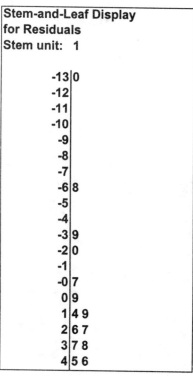

Chapter 12
Methods of Quality Improvement

12.1 Introduction

Chapter 12 introduces the topic of quality improvement to the reader. The main topic covered in the text involves using scatter plots to create control charts that monitor the outcomes of a statistical process. These control charts specify upper and lower limits on the plot inside which the process is expected to stay. By using control charts, the user has a valid statistical tool that enables him/her to identify when a process is decreasing in quality. PHStat offers three different types of control charts for the user. They are the $\bar{x}$, R, and p-charts. The following examples from *Statistics for Business and Economics* are solved with PHStat in this chapter.

Excel Companion		Statistics for Business and Economics	
Example	**Page**	**SBE**	**Page(s)**
12.1	154	Example 12.1/12.3	703-704, 716-717
12.3	157	Example 12.4	726-728

12.2 Constructing R-Charts and $\bar{x}$-Charts

The PHStat program allows for construction of the $\bar{x}$-chart only in conjunction with the R-chart. To create both of these charts within PHStat, **open** a new workbook and place the cursor in the upper left cell of the worksheet. **Click** on the **PHStat** menu at the top of the screen. **Select** the **Control Charts** option from the choices available and then select the **R & XBar Charts** option from those listed. You should open the R & XBar Charts menu that looks like the one shown in Figure 12.1.

Figure 5.1

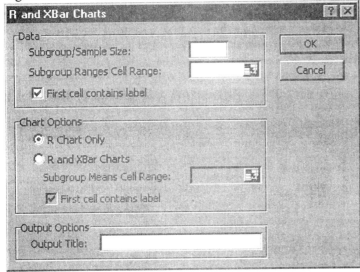

The user is required to enter the **Subgroup/Sample Size** and specify the location of the **Subgroup Ranges** (if only the R-Chart is desired) and the location of the **Subgroup Means** (if both the x̄-chart and R-charts are desired). The type of charts desired must be specified in the **Chart Options** section of the menu   An **Output Title** can be optionally selected if the user so desires. **Click OK** to finish. We illustrate with the next example.

Example 12.1: We combine Examples 12.1 found on pages 703-704 and Example 12.3 found on pages 716-717 of the *Statistics for Business and Economics* text.

Let's return to the paint-filling process described in Sections 12.2 and 12.3 of the text. Suppose instead of sampling 50 consecutive gallons of paint from the filling process to develop a control chart, it was decided to sample five consecutive cans once each hour for the next 25 hours. The sample data are presented in Table 12.1. This sampling strategy (rational subgrouping) was selected because several times a month the filling head in question becomes clogged. When that happens, the head dispenses less and less paint over the course of the day. However, the pattern of decrease is so irregular that minute-to-minute or even half-hour-to-half-hour changes are difficult to detect.

a.   Construct an x̄-chart for the process using the data below (from Example 12.1 in the text).
b.   Construct an R-chart for the process using the data below (from Example 12.2 in the text).

Table 12.1

Sample	Measurements					Mean	Range
1	10.0042	9.9981	10.001	9.9964	10.0001	9.99995	0.0078
2	9.995	9.9986	9.9948	10.003	9.9938	9.99704	0.0092
3	10.0028	9.9998	10.0086	9.9949	9.998	10.00082	0.0137
4	9.9952	9.9923	10.0034	9.9965	10.0026	9.998	0.0111
5	9.9997	9.9983	9.9975	10.0078	9.9891	9.99649	0.0195
6	9.9987	10.0027	10.0001	10.0027	10.0029	10.00141	0.0042
7	10.0004	10.0023	10.0024	9.9992	10.0135	10.00358	0.0143
8	10.0013	9.9938	10.0017	10.0089	10.0001	10.00116	0.0151
9	10.0103	10.0009	9.9969	10.0103	9.9986	10.00339	0.0134
10	9.998	9.9954	9.9941	9.9958	9.9963	9.99594	0.0039
11	10.0013	10.0033	9.9943	9.9949	9.9999	9.99874	0.009
12	9.9986	9.999	10.0009	9.9947	10.0008	9.99882	0.0062
13	10.0089	10.0056	9.9976	9.9997	9.9922	10.0008	0.0167
14	9.9971	10.0015	9.9962	10.0038	10.0022	10.00016	0.0076
15	9.9949	10.0011	10.0043	9.9988	9.9919	9.99822	0.0124
16	9.9951	9.9957	10.0094	10.004	9.9974	10.00033	0.0137
17	10.0015	10.0026	10.0032	9.9971	10.0019	10.00127	0.0061
18	9.9983	10.0019	9.9978	9.9997	10.0029	10.0013	0.0051
19	9.9977	9.9963	9.9981	9.9968	10.0009	9.99798	0.0127
20	10.0078	10.0004	9.9966	10.0051	10.0007	10.00212	0.0112
21	9.9963	9.999	10.0037	9.9936	9.9962	9.99764	0.0101
22	9.9999	10.0022	10.0057	10.0026	10.0032	10.00272	0.0058
23	9.9998	10.0002	9.9978	9.9966	10.006	10.00009	0.0094
24	10.0031	10.0078	9.9988	10.0032	9.9944	10.00146	0.0134
25	9.9993	9.9978	9.9964	10.0032	10.0041	10.00015	0.0077

Section 12.2: Constructing R-Charts and $\bar{x}$-Charts 155

Solution:

We begin by opening the data file SBE Example 12.1 to access the data. Next, **click** on the **PHStat** menu at the top of the screen. **Select** the **Control Charts** option from the choices available and then select the **R & XBar Charts** option from those listed. We enter the **Subgroup/Sample Size** of **5** (see Figure 12.2) and we choose the **R and XBar Charts** in the **Chart Options** section of the menu. We specify the location of both the **Subgroup Ranges** and the location of the **Subgroup Means.** We add an **Output Title** and **click OK** to finish. The charts are shown in Figures 12.3 and 12.4 below.

Figure 12.2

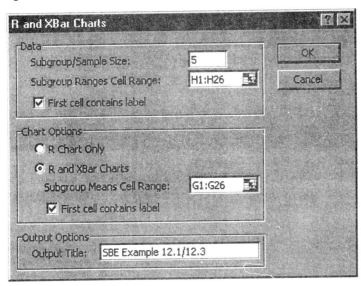

Figure 12.3

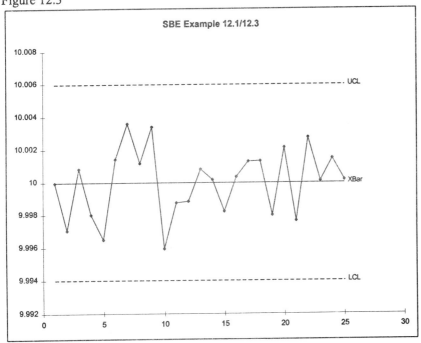

156 Chapter 12: Methods of Quality Improvement

Figure 12.4

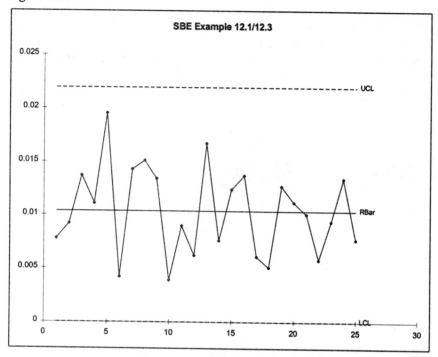

We refer the reader to Chapter 12 in the *Statistics for Business and Economics* text regarding information on how to read and interpret these control charts.

12.3 Constructing p-Charts

The PHStat program allows for construction of the p-chart for charting the proportion of defectives in a process. To create the p-chart within PHStat, **open** a new workbook and place the cursor in the upper left cell of the worksheet. **Click** on the **PHStat** menu at the top of the screen. **Select** the **Control Charts** option from the choices available and then select the **p Chart** option from those listed. You should open the pChart menu that looks like the one shown in Figure 12.5.

Figure 12.5

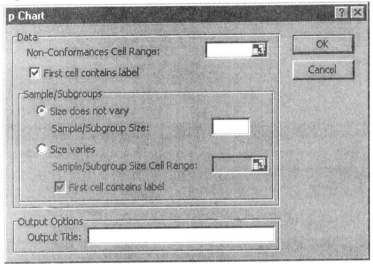

The user is required to enter the **Non-Conformances Cell Range** and specify the whether the **Samples/Subgroups** have equal or unequal sample sizes. If equal, the user is asked to specify the sample size. If unequal, the user must specify the location of these sample sizes. An **Output Title** can be optionally selected if the user so desires. **Click OK** to finish. We illustrate with the next example.

Example 12.3: We use Examples 12.4 found on pages 726-728 of the *Statistics for Business and Economics* text.

A manufacturer of auto parts is interested in implementing statistical process control in several areas within its warehouse operation. The manufacturer wants to begin with the order assembly process. Too frequently orders received by customers contain the wrong items or too few items.

For each order received, parts are picked from storage bins in the warehouse, labeled, and placed on a conveyor belt system. Since the bins are spread over a three-acre area, items that are part of the same order may be placed on different spurs of the conveyor belt system. Near the end of the belt system all spurs converge and a worker sorts the items according to the order they belong to. That information is contained on labels that were placed on the items by the pickers.

The workers have identified three errors that cause shipments to be improperly assembled: (1) pickers pick from the wrong bin, (2) pickers mislabel items, and (3) the sorter makes an error.

The firm's quality manager has implemented a sampling program in which 90 assembled orders are sampled each day and checked for accuracy. An assembled order is considered nonconforming (defective) if it differs in any way from the order placed by the customer. To date, 25 samples have been evaluated. The resulting data are shown in Table 12.2. Construct a p-chart for the order assembly operation.

Chapter 12: Methods of Quality Improvement

Table 12.2

Sample	Size	Defective Orders	Sample Proportion
1	90	12	0.13333
2	90	6	0.06667
3	90	11	0.12222
4	90	8	0.08889
5	90	13	0.14444
6	90	14	0.15556
7	90	12	0.13333
8	90	6	0.06667
9	90	10	0.11111
10	90	13	0.14444
11	90	12	0.13333
12	90	24	0.26667
13	90	23	0.25556
14	90	22	0.24444
15	90	8	0.08889
16	90	3	0.03333
17	90	11	0.12222
18	90	14	0.15556
19	90	5	0.05556
20	90	12	0.13333
21	90	18	0.20000
22	90	12	0.13333
23	90	13	0.14444
24	90	4	0.04444
25	90	6	0.06667
Totals	2,250	292	

Solution:

We begin by opening the data file SBE Example 12.4 to access the data. Next, **click** on the **PHStat** menu at the top of the screen. **Select** the **Control Charts** option from the choices available and then select the **p Chart** option from those listed. Since the data from this problem all contain samples of size 90, we select the **Size does not vary** Sample/Subgroup option and specify a **Sample/Subgroup Size** of 90 (see Figure 12.6). We specify the location of the **Non-Conformances** in the cell range location in the menu. We add an **Output Title** and **click OK** to finish. The chart is shown in Figure 12.7 below.

Section 12.3: Constructing p-Charts 159

Figure 12.6

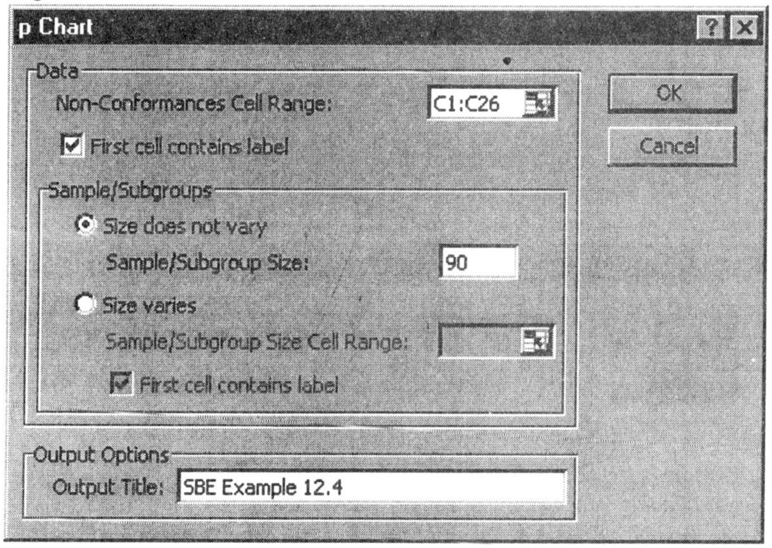

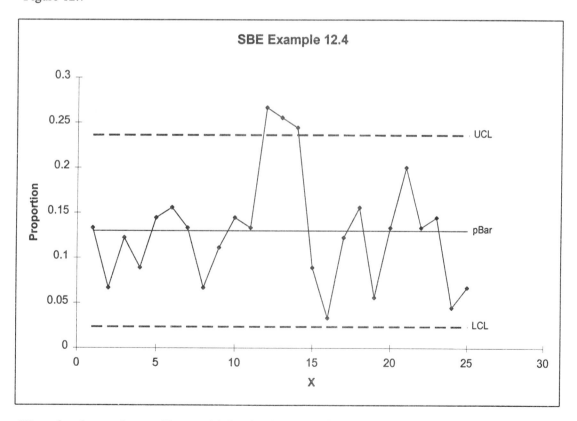

Figure 12.7

We refer the reader to Chapter 12 in the *Statistics for Business and Economics* text regarding information on how to read and interpret this control chart.

Technology Lab

The following exercises from the *Statistics for Business and Economics* text are given for you to practice the control chart procedures that are available within PHStat. Included with the exercises are the PHStat outputs that were generated to solve the problems.

12.73 Officials at Mountain Airlines are interested in monitoring the length of time customers wait in line to check in at their airport counter in Reno, Nevada. In order to develop a control chart, five customers were sampled each day for 20 days. The data, in minutes, are presented below (file SBE Exercise 12.73).

Sample	Waiting Times				
1	3.20	6.70	1.30	8.40	2.20
2	5.00	4.10	7.90	8.10	0.40
3	7.10	3.20	2.10	6.50	3.70
4	4.20	1.60	2.70	7.20	1.40
5	1.70	7.10	1.60	0.90	1.80
6	4.70	5.50	1.60	3.90	4.00
7	6.20	2.00	1.20	0.90	1.40
8	1.40	2.70	3.80	4.60	3.80
9	1.10	4.30	9.10	3.10	2.70
10	5.30	4.10	9.80	2.90	2.70
11	3.20	2.90	4.10	5.60	0.80
12	2.40	4.30	6.70	1.90	4.80
13	8.80	5.30	6.60	1.00	4.50
14	3.70	3.60	2.00	2.70	5.90
15	1.00	1.90	6.50	3.30	4.70
16	7.00	4.00	4.90	4.40	4.70
17	5.50	7.10	2.10	0.90	2.80
18	1.80	5.60	2.20	1.70	2.10
19	2.60	3.70	4.80	1.40	5.80
20	3.60	0.80	5.10	4.70	6.30

a. Construct an R-chart from these data.

d. Construct an $\bar{x}$-chart from these data.

PHStat Output

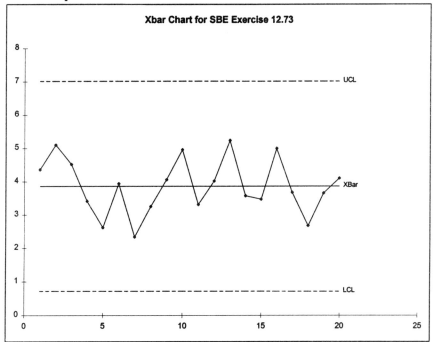

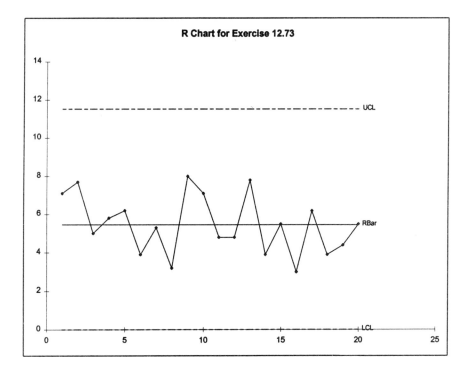

162   Chapter 12: Methods of Quality Improvement

12.74 A company called CRW runs credit checks for a large number of banks and insurance companies. Credit history information is typed into computer files by trained administrative assistants. The company is interested in monitoring the proportion of credit histories that contain one or more data entry errors. Based on her experience with the data entry operation, the director of the data processing unit believes that the proportion of histories with data entry errors is about 6%. CRW audited 150 randomly selected credit histories each day for 20 days. The sample data are presented below.

Sample	Sample Size	Histories with Errors
1	150	9
2	150	11
3	150	12
4	150	8
5	150	10
6	150	6
7	150	13
8	150	9
9	150	11
10	150	5
11	150	7
12	150	6
13	150	12
14	150	10
15	150	11
16	150	7
17	150	6
18	150	12
19	150	14
20	150	10

b.   Construct a p-chart for the data entry process.

PHStat Ouptut

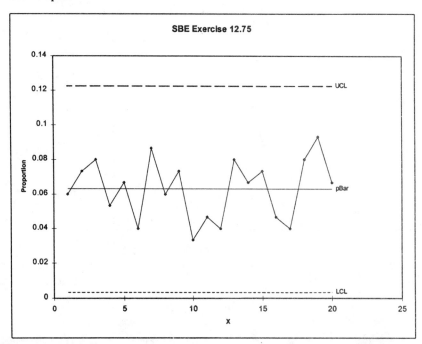

Chapter 13
Time Series: Descriptive Analyses, Models, and Forecasting

13.1 Introduction

Chapter 13 of the *Statistics for Business and Economics* text introduces the reader to the topic of Time Series Analysis. Descriptive analyses, time series modeling, and time series forecasting are the three main time series areas covered by the text.

While PHStat does not offer the user any time series techniques, Excel offers a variety of methods that enable the user to work with times series data. Many, like simple data manipulation, scatter plots, and regression analysis, have been encountered in one of the previous chapters of this manual. We reference these topics when we look at how they can be applied to times series data. Excel also offers times series tools that are not covered in the *Statistics for Business and Economics* text. Moving averages, Seasonal Indexes, and Cyclical Effects are topics offered within Excel but not covered in the text. We refer the reader to a more comprehensive text on time series analysis for information concerning these topics.

One other Excel data analysis tool, exponential smoothing, will be introduced in this chapter and needs further discussion here. The exponential smoothing technique that Excel offers differs from the exponential smoothing technique discussed in the book. There are many smoothing techniques available and the generic 'exponential' label is misleading. For the purpose of this manual, we do not utilize the Excel exponential smoothing data analysis tool. We, instead, use simple formula manipulation of the time series data to get the desired exponentially smoothed values that are discussed in the text.

There are several time series topics, however, that Excel is unable to provide assistance with. Section 13.5 of the text introduces the Holt-Winters forecasting model, an extension of the exponential smoothing topic covered in Section 13.2. Excel provides no data analysis tool to handle this more complicated smoothing model. In addition, the topic of measuring forecast accuracy (Section 13.6) has to Excel equivalent data analysis tool. And, finally, the Durbin-Watson test for autocorrelation is also not covered in Excel.

We will use the chapter examples that are given in the text to illustrate the model building and testing methods discussed above. The following examples from *Statistics for business and Economics* are solved with Microsoft Excel® in the Chapter:

Excel Companion		**Statistics for Business and Economics**	
Example	**Page**	**SBE**	**Page(s)**
13.1	164	Example 13.1	755
13.2	169	Example 13.2	757-758
13.3	170	Example 13.3	759-760
13.4	170	Example 13.4	767-768

164   Chapter 13: Time Series: Descriptive Analyses, Models, and Forecasting

13.2 Descriptive Analyses: Index Numbers

Index numbers are the most common techniques for characterizing the change in a business or economic data series over time. These indexes can be constructed in a variety of manner. The text introduces the reader to the simple index, the simple composite index, and two different weighted composite indexes (Laspeyres and Paasche indexes). We examine how Excel can be used to generate these indexes in the examples that follow.

Examples 13.1:   We use Example 13.1 found on page 755 in the *Statistics for Business and Economics* text.

One of the primary uses of index numbers is to characterize changes in stock prices over time. Stock market indexes have been constructed for many different types of companies and industries, and several composite indexes have been developed to characterize all stocks. These indexes are reported on a daily basis in the news media (e.g. Standard and Poor's 500 Stick Index and Dow Jones 65 Stock Index).

Consider the monthly closing prices (i.e., closing prices on the last day of each month) given in Table 13.1 for four high-technology company stocks listed on the New York Stock Exchange between 1995 and 1996. To see how this type of stock fared, construct a simple composite index using January 1998 as the base period. Graph the index, and comment on its implications.

Solution:

Index values are found by taking the value of he series at some point in time and dividing by the value of the series during the base period and then multiplying this ratio by 100. Simple Composite indexes use totals from several different time series as the values in the index ratio. For this problem, we will use Excel to calculate the sum of the monthly closing prices of the listed stocks, and then use Excel to simply divide this sum by the sum found in the base year. The simple composite index will be found by multiplying this ratio by 100.

We use very basic data manipulation techniques within Excel to generate the desired indexes. We begin by **opening** the Excel 97 file SBE Example 13.1. We will assume that the time series values are located in Columns C - F and Rows 2 - 25. The Column of totals should appear in Column G of the Excel worksheet. Note that Column H in the data sets has an abbreviated Month label. This column will be used to create the desired scatter plot. **Click** on the cell located in Row 2 Column 1. Enter **=(G2/$G$2)*100** in the cell. Excel should return the value 100 in the I2 cell. **Copy** the I2 cell to the cells located in Column I Rows 3 - 25 (e.g. I3 - I25). Compare the results returned by Excel (see Table 13.2) to those found on page 756 of the text. (Note: It is important that the denominator of the formula listed above include the dollar signs as this tells Excel to use the G2 cell as the base level in all subsequent calculations).

Table 13.1

Year	IBM	Motorola	Intel	Microsoft	Total
1998					
January	49.38	59.56	40.50	37.30	186.74
February	52.22	55.63	44.84	42.38	195.07
March	51.94	60.75	39.03	44.75	196.47
April	57.94	55.75	40.41	45.06	199.16
May	58.75	53.00	35.72	41.41	189.88
June	58.75	52.56	37.06	54.19	201.22
July	66.25	52.25	72.22	54.97	245.69
August	56.31	42.94	35.59	47.97	182.81
September	64.25	42.88	42.88	55.03	205.04
October	74.25	52.00	44.59	52.94	223.78
November	82.56	61.88	53.81	61.00	259.25
December	92.19	61.06	59.28	69.34	281.87
1999					
January	91.63	72.25	70.47	87.50	321.85
February	84.88	70.25	59.97	75.06	290.16
March	88.63	73.25	59.44	89.63	310.95
April	104.59	80.00	61.19	81.31	327.09
May	116.00	82.81	54.06	81.81	334.68
June	129.25	94.75	59.50	90.19	373.69
July	125.69	91.25	69.00	85.81	371.75
August	124.56	92.25	82.19	92.56	391.56
September	121.00	88.00	74.31	90.56	373.87
October	98.25	97.31	77.44	92.56	365.56
November	103.06	116.68	76.69	91.05	387.48
December	107.88	14725	8231	116.75	454.19

Table 13.2

Year	Month	IBM	Motorola	Intel	Microsoft	Total	Index
1998	January	49.38	59.56	40.50	37.30	186.74	100.0
	February	52.22	55.63	44.84	42.38	195.07	104.5
	March	51.94	60.75	39.03	44.75	196.47	105.2
	April	57.94	55.75	40.41	45.06	199.16	106.7
	May	58.75	53.00	35.72	41.41	189.88	101.7
	June	58.75	52.56	37.06	54.19	201.22	107.8
	July	66.25	52.25	72.22	54.97	245.69	131.6
	August	56.31	42.94	35.59	47.97	182.81	97.9
	September	64.25	42.88	42.88	55.03	205.04	109.8
	October	74.25	52.00	44.59	52.94	223.78	119.8
	November	82.56	61.88	53.81	61.00	259.25	138.8
	December	92.19	61.06	59.28	69.34	281.87	150.9
1999	January	91.63	72.25	70.47	87.50	321.85	172.4
	February	84.88	70.25	59.97	75.06	290.16	155.4
	March	88.63	73.25	59.44	89.63	310.95	166.5
	April	104.59	80.00	61.19	81.31	327.09	175.2
	May	116.00	82.81	54.06	81.81	334.68	179.2
	June	129.25	94.75	59.50	90.19	373.69	200.1
	July	125.69	91.25	69.00	85.81	371.75	199.1
	August	124.56	92.25	82.19	92.56	391.56	209.7
	September	121.00	88.00	74.31	90.56	373.87	200.2
	October	98.25	97.31	77.44	92.56	365.56	195.8
	November	103.06	116.68	76.69	91.05	387.48	207.5
	December	107.88	14725	8231	116.75	454.19	243.2

The scatter plot can be drawn in Excel using the Chart Wizard that was introduced in Chapter 2 of this manual. Click on the **Chart Wizard** icon that is located at the top of the Excel worksheet (see Figure 13.1). Highlight the **Line** Chart type and select the desired **Chart sub-type** that you desire. Click **Next**.

Figure 13.1

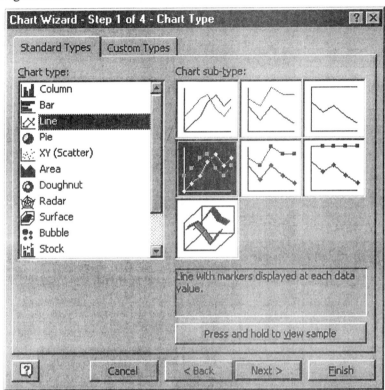

Enter the range of the data to be charted in the **Data Range** area of the menu (see Figure 13.2). Specify whether the data are listed in rows or **columns** by indicating the appropriate selection. Click **Next**.

The remaining Chart Wizard menus allow the user to customize the way the resulting line plot appears. We refer you the Chapter 2 of this manual to find a more thorough discussion of the options available within Excel. See Figure 13.3 for the plot created with Excel for the data of this example. Compare the Excel plot to the one found on page 752 of the *Statistics for Business and Economics* text.

168 Chapter 13: Time Series: Descriptive Analyses, Models, and Forecasting

Figure 13.2

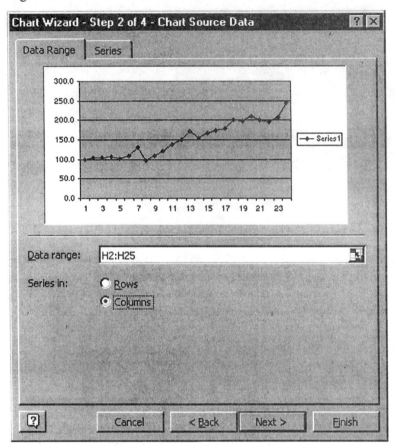

Figure 13.3

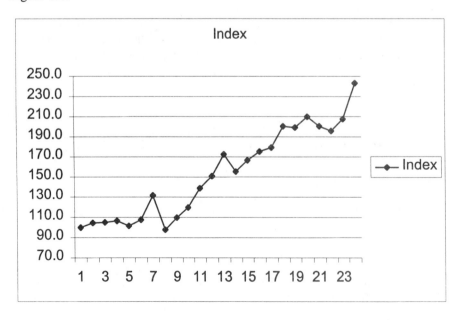

Section 13.2: Descriptive Analyses: Index Numbers

The simple composite index of Example 13.1 is found by summing the values of several times series and dividing by the sum from a base year. Each of the time series is given an equal weight in the simple composite index. Several different types of weighted composite indexes exist and the text discusses two such weighted composite indexes, Laspeyres and Paasche Indexes. We illustrate the Laspeyres Index below.

Example 13.2: We use Example 13.2 found on pages 757-758 in the *Statistics for Business and Economics* text.

The January prices for the four high technology company stocks are given in Table 13.3. Suppose that, in January 1998, an investor purchased the quantities shown in the table. [*Note*: Only January prices and quantities are used to simplify the example. The same methods can be applied to calculate the index for other months.] Calculate the Laspeyres index for the investor's portfolio of high-technology stocks using January 1998 as the base period.

Table 13.3

	IBM	Motorola	Intel	Microsoft
Shares Purchased	500	100	100	1000
January 1998 Price	49.38	59.56	40.5	37.3
January 1999 Price	91.63	72.25	70.47	87.5

Solution:

The first step in finding the Laspeyres indexes is to calculate the weighted price totals for each time period, using the January 1998 quantities as the weights. We begin by **opening** the Excel 97 file SBE Example 13.2. We will assume the weights are located in Column B - E, Row 2 in the worksheet and the time series values are located in columns B - E, Rows 3 - 4. **Click** on the cell located in Column F, Row 3. Enter **=$B$2*B3+$C$2*C3+$D$2*D3+$E$2*E3** in the cell. Make certain to use the "$" signs whenever you identify the cell locations of the weights. **Copy** the contents of Cell F3 to all the other cells in which weighted price totals are desired (e.g., Cell F4 in this example). Excel should return the weighted price totals shown below in Table 13.4.

The next step is to calculate the Laspeyres indexes using January 1998 as the base period. Click on the cell located in Column G, Row 3. Enter **=(F3/$F$3)*100** in the cell. **Copy** the contents of Cell G3 to all the other cells in which the Laspeyres indexes are desired (e.g., Cell G4 in this example). Excel should return the Laspeyres indexes shown below in Table 13.4.

Table 13.4

	IBM	Motorola	Intel	Microsoft	Weighted Total	Lapeyres Index
Shares Purchased	500	100	100	1000		
January 1998 Price	49.38	59.56	40.5	37.3	71,996	100.0
January 1999 Price	91.63	72.25	70.47	87.5	147,587	205.0

Compare the results returned by Excel to those found on page 758 of the text.

As we have seen in the last example, the Laspeyres Index uses the quantities of the base period as the weights for all of the average and index values that are calculated. In some instances, it is preferred to use the quantities from the current time period as the weights in the index calculations. One

170 Chapter 13: Time Series: Descriptive Analyses, Models, and Forecasting

method of achieving these indexes is to use the Paasche index. An example of the Paasche Index follows.

Example 13.3: We use Example 13.3 found on pages 741-742 in the *Statistics for Business and Economics* text.

The January prices and volumes (actual quantities purchased) in thousands of shares for the four high-technology company stocks are shown for 1998 and 1999 in Table 13.5. Calculate and interpret the Paasche index, using January 1998 as the base period.

Table 13.5

	IBM		Motorola		Intel		Microsoft	
	Price	Volume	Price	Volume	Price	Volume	Price	Volume
January 1998 Price	49.38	206.2	59.56	57.8	40.5	679.4	37.3	743
January 1999 Price	91.63	246.7	72.25	84.3	70.47	697.2	87.5	736

Solution

The first step in finding the Paasche indexes in to calculate the weighted price totals for each time period, using the current time period quantities as the weights. We begin by **opening** the Excel 97 file SBE Example 13.3. **Click** on the specified cell in the worksheet and enter:

For Q_{1999}, P_{1999}: **=B4*C4+D4*E4+F4*G4+H4*I4** in cell A6
For Q_{1999}, P_{1998}: **=B3*C4+D3*E4+F3*G4+H3*I4** in cell A7

For the Paasche index for 1999, enter **=(A6/A7)*100** in cell A9. Excel should return the Paasche index shown below in Table 13.6. Compare the results returned by Excel to those found on page 759 of the text.

Table 13.6

Q_{1999}, P_{1999}	142253.73
Q_{1999}, P_{1998}	72903.54
Paasche Index	195.1

Using indexes is just one method of describing time series data. We turn our attention now towards the use of exponential smoothing as a method to describe times series data.

13.3 Exponential Smoothing

A second method of describing time series data involves averaging past and current data together. The goal of the averaging is to reduce the volatility that is inherent to any time series. Exponential smoothing is one type of averaging method that allows the user to select the amount of weight given to the past and to the present data. This weight, known as the smoothing constant, is selected to be a number between 0 and 1. The larger the value of the smoothing constant, the more weight is given to the most current data value from the time series. We demonstrate with the following example.

Example 13.4: We use Example 13.4 found on pages 767-768 in the *Statistics for Business and Economics* text.

Consider the IBM common stock price from January 1998 to December 1999, shown in Table 13.7. Create the exponentially smoothes series using $w = .5$, and plot both series.

Table 13.7

1998	IBM	1999	IBM
January	49.38	January	91.63
February	52.22	February	84.88
March	51.94	March	88.63
April	57.94	April	104.59
May	58.75	May	116
June	57.41	June	129.25
July	66.25	July	125.69
August	56.31	August	124.56
September	64.25	September	121
October	74.25	October	98.25
November	82.56	November	103.06
December	92.19	December	107.88

Solution:

We begin by **opening** Excel 97 file SBE Example 13.4. We will assume that the IBM stock prices appear in Column B, Rows 2 - 37 of the worksheet. To find the first exponentially smoothed value, enter =B2 in cell C2. To find the second exponentially smoothed value, we enter =.5*B3+(1-.5)*C2 in cell C3. We use the value of .5 in this equation because the smoothing constant for this problem is $w = .5$. Copy the formula of cell C3 to the cells C4 through C37 to obtain the rest of the exponentially smoothed values. Table 13.8 shows the values of the original time series as well as the exponentially smoothed values. Compare these values to the ones found on page 768 of the text.

Table 13.8

1998	IBM	Smoothed	1999	IBM	Smoothed
January	49.38	49.38	January	91.63	87.67
February	52.22	50.80	February	84.88	86.28
March	51.94	51.37	March	88.63	87.45
April	57.94	54.66	April	104.59	96.02
May	58.75	56.70	May	116.00	106.01
June	57.41	57.06	June	129.25	117.63
July	66.25	61.65	July	125.69	121.66
August	56.31	58.98	August	124.56	123.11
September	64.25	61.62	September	121.00	122.06
October	74.25	67.93	October	98.25	110.15
November	82.56	75.25	November	103.06	106.61
December	92.19	83.72	December	107.88	107.24

Figure 13.4

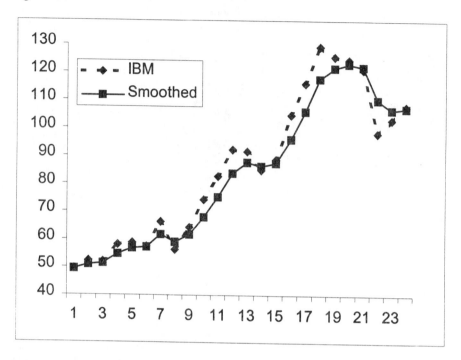

A plot of both the original times series the exponentially smoothed series is shown above in Figure 13.4. Compare this chart to the one shown on page 768 of the text. Note that any value of a smoothing constant can be used by changing the appropriate values in the formula used to calculate the exponentially smoothed values of the series. The exponentially smoothed values will change depending whether the constant places more weight on the current value of the past values of the time series.

Section 13.5 introduces the reader to a second type of smoothing process, the Holt-Winters smoothing model, as a method for forecasting values of a times series. The book gives the necessary formulas that can be used within Excel to find the Holt-Winters values. We leave this formula manipulation to the user. We recommend using the last example as a guide and substitute the Holt-Winters formulas for those used in the example. Use Example 13.6 from the text (page 775) to validate your work.

13.4 Using Regression to Model Time Series Data

Sections 13.7 and 13.8 introduce the reader to using regression models to model the linear trend and seasonal variation in time series data. The linear trend component can be modeled by using a measure of the time period as a quantitative variable in the regression. For example, the model $E(Y_t) = \beta_0 + \beta_1 t$ can be used to forecast the value of a time series at time period t. This model would assume that the time series values increase linearly over time.

While the linear model discussed above works in some applications, many times series data are affected by some sort of cyclical, or seasonal, influences. This cyclical variation can be modeled in regression using the qualitative variables discussed in Chapter 11. The seasonal component must be identified and explained using qualitative indicator variables. For example, the model

$E(Y_t) = \beta_0 + \beta_1 t + \beta_2 Q_1 + \beta_3 Q_2 + \beta_4 Q_3$ could be used to include both a linear trend (modeled with the quantitative time period variable, t) and a seasonal component (modeled with the three indicator variables Q_1, Q_2, and Q_3). The Q's in this model would be appropriate if the time series was influenced by some quarterly effect. For a monthly effect, the model would need to include eleven indicator variables.

We will not fit regression models to the time series data of Chapter 13. We remind the user that Excel requires the data set to include all of the independent variables to be included in the regression model. The variables must be in adjacent columns of the Excel worksheet. We refer the user to Chapter 10 - 11 of this manual to review how to fit regression models within Excel. The regression models fit will yield estimates to the values of the time series data of Chapter 13.

Chapter 14
Design of Experiments and Analysis of Variance

14.1 Introduction

Chapter 14 introduces the topics of design of experiments and analysis of variance (ANOVA) to the reader. The concept of the designed experiment is explained and the completely randomized and factorial designs are covered in the text. The goal of analysis of variance is to identify factors that contribute information to the response variable of interest. The combination of levels of the various factors are called treatments and the analysis of variance procedures discussed in the text attempt to detect differences in the mean response variable for the various treatments. Once detected, the text presents several methods of comparing the multiple means of the experiment. The final topic covered in the text relates the designed experiments of this chapter to the regression models of chapters 10 and 11.

PHStat does not offer any analysis of variance procedures. However, Excel 97 offers two analysis of variance procedures that can be used for the completely randomized and factorial designs. These data analysis tools are very easy to implement. In addition, the regression data analysis tool that we examined in Chapters 10 and 11 can also be used to fit the ANOVA models of this chapter. Excel does not, however, offer a follow-up tool to compare treatment means that have been determined to differ. The Excel user must take the summary results from the two analyses and calculate the multiple comparison procedures by hand.

We will use the chapter examples that are given in the text to illustrate the model building and testing methods discussed above. The following examples from *Statistics for Business and Economics* are solved with Microsoft Excel® in this chapter.

Excel Companion		Statistics for Business and Economics	
Example	Page	SBE	Page(s)
14.1	175	Example 14.3	826-827
14.2	178	Example 14.6	853-855

14.2 The Completely Randomized Design

The goal of analysis of variance is to compare the mean responses of the various treatments in an experimental design, where the treatments are the combinations of the levels of all the factors involved in the design. The simplest of all experimental designs involves using a single factor to compare values of a response variable. Since there is only one factor in the design, the various levels of the factor are the treatments in the design. The goal is to compare the means of the response variable for those treatments. This experimental design is the completely randomized design and can be analyzed in Excel using the Anova: Single Factor data analysis tool. We illustrate with the following example.

Example 14.1: We use Example 14.3 found on pages 826-827 in the *Statistics for Business and Economics* text.

176 Chapter 14: Design of Experiments and Analysis of Variance

Suppose the United States Golf Association (USGA) wants to compare the mean distances associated with four different brands of golf balls when struck with a driver. A completely randomized design is employed, with Iron Byron, the USGA's robotic golfer, using a driver to hit a random sample of 10 balls of each brand in a random sequence. The distance is recorded for each hit, and the results are shown in Table 15.1, organized by brand.

a. Set up the test to compare the mean distances for the four brands. Use $\alpha = .10$.
b. Use Excel to obtain the test statistic and p-value. Interpret the results.

Table 14.1

Brand A	Brand B	Brand C	Brand D
251.2	263.2	269.7	251.6
245.1	262.9	263.2	248.6
248.0	265.0	277.5	249.4
251.1	254.5	267.4	242.0
260.5	264.3	270.5	246.5
250.0	257.0	265.5	251.3
253.9	262.8	270.7	261.8
244.6	264.4	272.9	249.0
254.6	260.6	275.6	247.1
248.8	255.9	266.5	245.9

Solution:

We begin by **opening** the Excel 97 file SBE Example 14.3. Click on the **Tools** menu located at the top of the Excel worksheet. Select the **Data Analysis** option from within the Tools menu and highlight the **Anova: Single Factor** option (see Figure 14.1). Click **OK**.

Figure 14.1

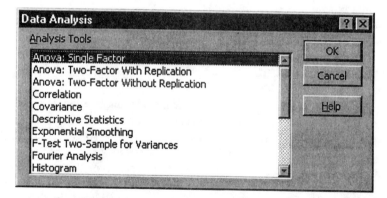

Either type or click the rows and columns where the input data is located and enter this information into the **Input Range** area of the Anova: Single Factor menu (see Figure 14.2). Select the manner in which the data is grouped (**Columns** or **Rows**) and give a level of significance in the **Alpha** cell of the menu (e.g., .10). Specify the location of the computed output by selecting either the **Output Range**, New Worksheet Ply, or New Workbook option, and entering the corresponding **cell** or name. Click **OK**.

Section 14.2: The Completely Randomized Design

Figure 14.2

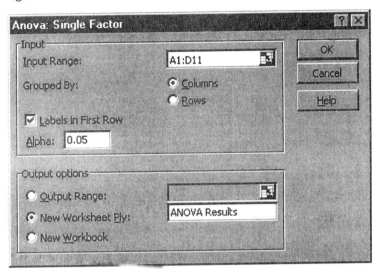

The ANOVA printout generated for the completely randomized design has two main components (see Table 14.2). The first component is a statistical summary of the various treatments in the analysis. For each of the four brands of balls, Excel gives some summary information concerning the distances achieved by each. This information will be more useful after studying the multiple comparison of means material in Section 14.3.

Table 14.2

Anova: Single Factor							
SUMMARY							
Groups	Count	Sum	Average	Variance			
Brand A	10	2507.8	250.78	22.42177778			
Brand B	10	2610.6	261.06	14.94711111			
Brand C	10	2699.5	269.95	20.25833333			
Brand D	10	2493.2	249.32	27.07288889			
ANOVA							
Source of Variation	SS	df	MS	F	P-value	F crit	
Between Groups	2794.389	3	931.462917	43.98874592	3.97311E-12	2.242607877	
Within Groups	762.301	36	21.1750278				
Total	3556.69	39					

The second component is called the analysis of variance table and is where the pertinent testing information will be found. To test whether the mean distances of the four means differ, we use the test statistic and p-value found in the Brand row of the printout (labeled on the Excel printout as the Between Groups row). We see that the test statistic is F = 43.9887 and the p-value is $p \approx 0$. Compare these values to the values found in the SAS printout found on pages 827 of the text. We refer you to the text for further information regarding the interpretation of these values.

178 Chapter 14: Design of Experiments and Analysis of Variance

14.3 The Factorial Design

The next step in the experimental design process is to add a second factor to the design. One possible design that results is the factorial design. In Excel, the data analysis procedure that should be used is the Anova: Two-Factor With Replication procedure. This procedure allows both the factors to be analyzed as well as the interaction between them. We illustrate its use with the following example.

Example 14.2: We use Example 14.6 found on pages 853-855 in the *Statistics for Business and Economics* text.

Suppose the United States Golf Association (USGA) tests four different brands (A, B, C, D) of golf balls and two different clubs (driver, five-iron) in a completely randomized design. Each of the eight Brand-Club combinations (treatments) is randomly and independently assigned to four experimental units, each experimental unit consisting of a specific position in the sequence of hits by Iron Byron. The distance response is recorded for each of the 32 hits, and the results are shown in Table 14.3.

a. Use Excel to partition the Total Sum of Squares into the components necessary to analyze this 4x2 factorial experiment.
b. Follow the steps for analyzing a two-factor factorial experiment and interpret the results of your analysis. Use $\alpha = .10$ for the tests you conduct.

Table 14.3

		BRAND			
		A	B	C	D
	DRIVER	226.4	238.3	240.5	219.8
	DRIVER	232.6	231.7	246.9	228.7
	DRIVER	234.0	227.7	240.3	232.9
CLUB	DRIVER	220.7	237.2	244.7	237.6
	FIVE-IRON	163.8	184.4	179.0	157.8
	FIVE-IRON	179.4	180.6	168.0	161.8
	FIVE-IRON	168.6	179.5	165.2	162.1
	FIVE-IRON	173.4	186.2	156.5	160.3

Solution:

We begin by **opening** the Excel 97 file SBE Example 14.6. Click on the **Tools** menu located at the top of the Excel worksheet. Select the **Data Analysis** option from within the Tools menu and highlight the **Anova: Two Factor With Replication** option (see Figure 14.3). Click **OK**.

Figure 14.3

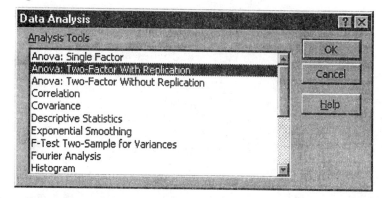

Section 14.3: The Factorial Design

Either type or click the rows and columns where the input data is located and enter this information into the **Input Range** area of the Anova: Two Factor With Replication menu (see Figure 14.4). Include the labels for the columns and rows of the factor when inputting the data range. Note that each Brand of golf ball includes four rows of data for each of the two Clubs tested. Enter the number of rows (e.g., 4) in the **Rows per Sample** area of the menu. Give a level of significance in the **Alpha** cell of the menu (e.g., .10). Specify the location of the computed output by selecting either the **Output Range,** New Worksheet Ply, or New Workbook option, and entering the corresponding **cell** or name. Click **OK**.

Figure 14.4

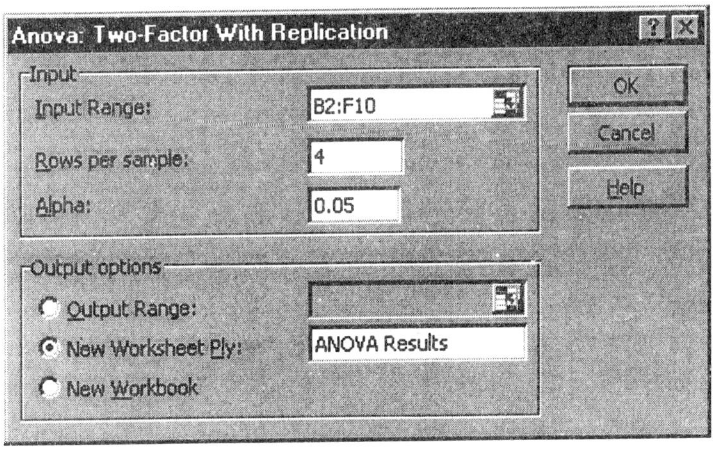

The ANOVA printout generated for the factorial design has two main components (see Table 14.4). The first component is a statistical summary of the various treatments in the analysis. For each of the eight Brand-Club treatments, Excel gives some summary information concerning the distances achieved by each. This information can be used when comparing treatment means similar to the methods used in Section 14.3.

The second component is the analysis of variance table and is where the pertinent testing information will be found. This is where the sums of squares are petitioned into the various components and where the interaction and individual factor test statistics and p-values are located. Compare this printout to the one found on page 854 of the text.

The first test of interest to the USGA is to determine if interaction exists between the Club and Brand factors in the experiment. We use the test statistic (t = 7.452435) and the p-value (p = 0.001079) found in the interaction row of the analysis of variance table. Refer to the text for more information concerning the interpretation of these values and the follow-up analysis that is necessary for factorial designs.

We note here one drawback associated with the Excel analysis for the completely randomized and factorial designs in the analysis of variance experiments. Excel offers the appropriate analyses for determining when differences exist between the treatment means for each of these two experimental designs, but does not offer any method to determine where the specific differences exists. Both the SAS™ and MINITAB™ software packages offer options for conducting the multiple comparison procedures that enable the user to conduct the appropriate follow-up analysis for both the completely randomized and factorial designs. Consult Section 14.3 and the references at the end of the text for more information on this topic.

Table 14.4

Anova: Two-Factor With Replication					
SUMMARY	A	B	C	D	Total
DRIVER					
Count	4	4	4	4	16
Sum	913.7	934.9	972.4	919	3740
Average	228.425	233.725	243.1	229.75	233.75
Variance	37.429167	24.46917	10.53333	57.21667	61.07067
FIVE-IRON					
Count	4	4	4	4	16
Sum	685.2	730.7	668.7	642	2726.6
Average	171.3	182.675	167.175	160.5	170.4125
Variance	44.52	9.929167	86.1225	3.86	98.19183
Total					
Count	8	8	8	8	
Sum	1598.9	1665.6	1641.1	1561	
Average	199.8625	208.2	205.1375	195.125	
Variance	967.48268	759.3429	1688.454	1396.336	

ANOVA						
Source of Variation	SS	df	MS	F	P-value	F crit
Sample	32093.111	1	32093.11	936.7516	9.63E-21	2.927116
Columns	800.73625	3	266.9121	7.790779	0.00084	2.32739
Interaction	765.96125	3	255.3204	7.452435	0.001079	2.32739
Within	822.24	24	34.26			
Total	34482.049	31				

14.4 Using Regression Analysis for ANOVA

The analysis of variance designs of this chapter can be thought of as being composed of a quantitative dependent variable (called the response variable) and one (completely randomized design) or two (factorial design) qualitative independent variables (called factors). The regression modeling techniques discussed in Chapter 11 can easily be applied to the analysis of variance experiments of this chapter. The key to using regression analysis for ANOVA lies in defining the independent variables correctly. We refer the reader to the material regression topics covered in Chapter 11 of this manual. The PHStat regression techniques allow the user to easily model these analysis of variance designs with the regression models from Chapter 11.

Chapter 15
Nonparametric Statistics

15.1 Introduction

Chapter 15 introduces the reader to the topic of nonparametric statistics and gives the reader several different examples of methods to analyze data using the nonparametric techniques. Two of these techniques are available within the PHStat software program. PHStat offers both the Wilcoxon Rank Sum test for independent samples and the Kruskal-Wallis H-Test for a Completely Randomized Design.

Both techniques require the user to have access to the data that is being analyzed. After specifying the location of the data, PHStat offers the user an easy method for analyzing the data using these two procedures. The following examples from *Statistics for Business and Economics* are solved using PHStat in this chapter:

Excel Companion		Statistics for Business and Economics	
Example	Page	SBE	Page(s)
15.1	182	Table 15.7	911-914

15.2 The Wilcoxon Rank Sum Test for Independent Samples

The *Statistics for Business and Economics* text offers the Wilcoxon Rank Sum technique for comparing two population means with independent samples. The only procedure available in PHStat is the large sample approximation of the Wilcoxon Rank Sum technique. PHStat allows the user to perform this test of hypothesis when the data has been collected using two random, independent samples. To use the test of hypothesis tool within PHStat, **open** a new workbook and place the cursor in the upper left cell of the worksheet. **Click** on the **PHStat** menu at the top of the screen. **Select** the **Two-Sample Tests** option from the choices available and then select the **Wilcoxon Rank Sum Test** option from those listed. You should open the Wilcoxon Rank Sum Test menu that looks like the one shown in Figure 15.1.

Figure 15.1

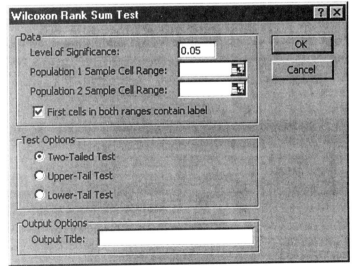

182 Chapter 15: Nonparametric Statistics

The user is required to enter the **Level of Significance**, the location of the sampled data in the **Cell Range** area of the menu, and the direction of the test in the **Test Options.** An **Output Title** can be optionally selected if the user so desires. **Click OK** to finish. PHStat gives both the test statistic and the p-value for the test to the user. None of the examples in the text utilize large independent samples, so this procedure is not illustrated here. See the Technology Lab at the end of the chapter for an example of this procedure.

15.3 The Kruskal-Wallis H-Test for the Completely Randomized Design

The *Statistics for Business and Economics* text offers the Kruskal-Wallis H-Test for comparing population means from samples collected utilizing the completely randomized design. To use the test of hypothesis tool within PHStat, **open** a new workbook and place the cursor in the upper left cell of the worksheet. **Click** on the **PHStat** menu at the top of the screen. **Select** the **c-Sample Tests** option from the choices available and then select the **Kruskal-Wallis Rank Test** option from those listed. You should open the Kruskal-Wallis Rank Test menu that looks like the one shown in Figure 15.2.

Figure 15.2

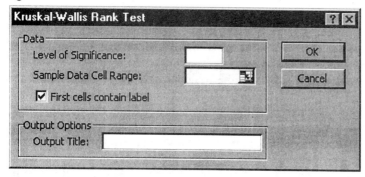

The user is required to enter the **Level of Significance** and the location of the sampled data in the **Cell Range** area of the menu. An **Output Title** can be optionally selected if the user so desires. **Click OK** to finish. PHStat gives both the test statistic and the p-value for the test to the user. We illustrate this technique with the following example.

Example 15.1: Use the data from Table 15.7 of the Statistics for Business and Economics text to determine if the distributions for the number of beds available at the three hospitals differ. Use the Kruskal-Wallis H-test and a .05 level of significance.

Data:

Hospital 1	Hospital 2	Hospital 3
6	34	13
38	28	35
3	42	19
17	13	4
11	40	29
30	31	0
15	9	7
16	32	33
25	39	18
5	27	24

Section 15.3: The Kruskal-Wallis H-Test for the Completely Randomized Design

Solution:

We first must open the data set **SBE Table 15.7** (included on the enclosed data disk). After accessing the Kruskal-Wallis Rank Test menu, we **specify a .05 Level of Significance** and the **location** of the Sample Data Cell Range (See Figure 15.3). We opt to **add** the **title, SBE Table 15.7**. We **click OK** to finish. The output generated by PHStat is shown in Table 15.1.

Figure 15.2

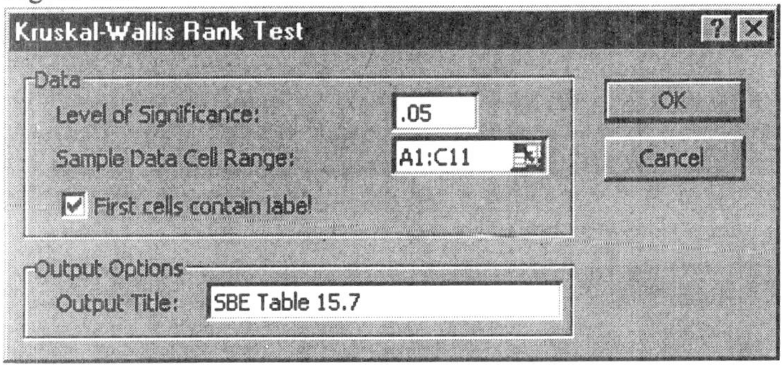

Table 15.1

SBE Table 15.7	
Level of Significance	0.05
Group 1	
Sum of Ranks	120
Sample Size	10
Group 2	
Sum of Ranks	210.5
Sample Size	10
Group 3	
Sum of Ranks	134.5
Sample Size	10
Sum of Squared Ranks/Sample Size	7680.05
Sum of Sample Sizes	30
Number of groups	3
H Test Statistic	6.097419
Critical Value	5.991476
p-Value	0.04742
Reject the null hypothesis	

Compare the value of the test statistic and the p-value shown here to the values shown on the SAS printout shown on page 914 of the text.

Technology Lab

The following exercises from the *Statistics for Business and Economics* text are given for you to practice the nonparametric procedures that are available within PHStat. Included with the exercises are the PHStat outputs that were generated to solve the problems.

15.67 Weevils cause millions of dollars worth of damage each year to cotton crops. Two chemicals (A and B) designed to control weevil populations were applied, one to each of two cotton fields of cotton. After three months, 10 plots of equal size were randomly selected within each field and the percentage of cotton plants with weevil damage was recorded for each. Do the data in the accompanying table provide sufficient evidence to indicate a difference in location among the distributions of damage rates corresponding to the two chemical treatments. Use $\alpha = .05$.

A	B
10.8	22.3
15.6	19.5
19.2	18.6
17.9	24.3
18.3	19.9
9.8	20.4
16.7	23.6
19	21.2
20.3	19.8
19.4	22.6

PHStat Output

SBE Exercise 15.67	
Level of Significance	0.05
Population 1 Sample	
Sample Size	10
Sum of Ranks	62
Population 2 Sample	
Sample Size	10
Sum of Ranks	148
Total Sample Size n	20
T1 Test Statistic	62
T1 Mean	105
Standard Error of T1	13.22876
Z Test Statistic	-3.25049
Two-Tailed Test	
Lower Critical Value	-1.95996
Upper Critical Value	1.959961
p-value	0.001152
Reject the null hypothesis	

15.72 A savings and loan association is considering three locations in a large city as potential office sites. The company has hired a marketing firm to compare the incomes of people living in the area surrounding each site. The market researchers interview 10 households chosen at random in each area to determine the type of job, length of employment, etc., of those in the households who work. This information enables them to estimate the annual income of each household, as shown in the table (in thousands of dollars).

Site 1	Site 2	Site 3
34.3	39.3	34.5
35.5	45.5	29.3
32.1	50.2	37.2
29.3	72.1	33.2
40.5	48.6	32.6
36.2	42.2	38.3
43.5	103.5	43.3
34.7	47.9	36.7
38.0	41.2	40.0
35.1	44.0	35.2

b. Use the appropriate nonparametric test to compare the treatments. Specify the hypotheses and interpret the results in terms of this experiment. Use $\alpha = .05$.

PHStat Output

SBE Exercise 15.72	
Level of Significance	0.05
Group 1	
Sum of Ranks	107.5
Sample Size	10
Group 2	
Sum of Ranks	247
Sample Size	10
Group 3	
Sum of Ranks	110.5
Sample Size	10
Sum of Squared Ranks/Sample Size	8477.55
Sum of Sample Sizes	30
Number of groups	3
H Test Statistic	16.38774
Critical Value	5.991476
p-Value	0.000276
Reject the null hypothesis	

Chapter 16
Categorical Data Analysis

16.1 Introduction

Chapter 3 introduces the topic of categorical data analysis to the reader. Both the one-way and two-way analyses are discussed in the text. PHStat allows the user to work with the two-way analyses found in the contingency tables covered in section 16.3 of the text. We illustrate the PHStat two-way analysis by using the following example from the text:

Excel Companion		Statistics for Business and Economics	
Example	**Page**	**SBE**	**Page(s)**
16.1	187	Example 16.3	950-951

16.2 Testing Categorical Probabilities: Two Way Table

Chapter 3 introduced the reader to the idea of presenting descriptive results of collected data in a tabular form. In Chapter 16, we now take a look at a technique that allows us to determine if the outcomes of these two variables are dependent upon one another. The two-way analysis of data is available in PHStat through using the **Chi-Square Test** found in the **c-Sample Test** option of the **PHStat** menu. The general Chi Square Test menu is shown in Figure 16.1. The user is required to initially enter a **Level of Significance**

Figure 16.1

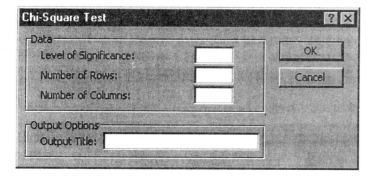

for the desired test, and the **Number of Rows and Columns** for the data that is being analyzed. An optional **Output Title** is available, and **OK** is **selected** to finish. The selections made in this menu will create a worksheet that the user can customized to generate the desired test for independence between two variables. We illustrate in the following example.

Example 16.1 Use example 16.3 found on pages 950-951 of the *Statistics for Business and Economics* text.

188 Chapter 16: Categorical Data Analysis

A large brokerage firm wants to determine whether the service it provides to affluent clients differs from the service it provides to lower-income clients. A sample of 500 clients is selected, and each client is asked to rate his or her broker. The results are shown below in Table 16.1.

Table 16.1

		Clients Income			
		Under $30,000	$30-60,000	Over $60,000	Totals
	Outstanding	48	64	41	153
Broker	Average	98	120	50	268
Rating	Poor	30	33	16	79
	Totals	176	217	107	500

a. Test to determine whether there is evidence that broker rating and customer income are independent. Use $\alpha = .10$.

Solution:

In the Chi-Square Test menu, we enter a **Level of Significance** of **.10** (See Figure 16.2), a **Number of Rows** of **3**, and a **Number of Columns** of **3**. We enter an **Output Title** and **click OK**. These selections create a worksheet with a 3x3-table template (See Figure 16.3)

Figure 16.2

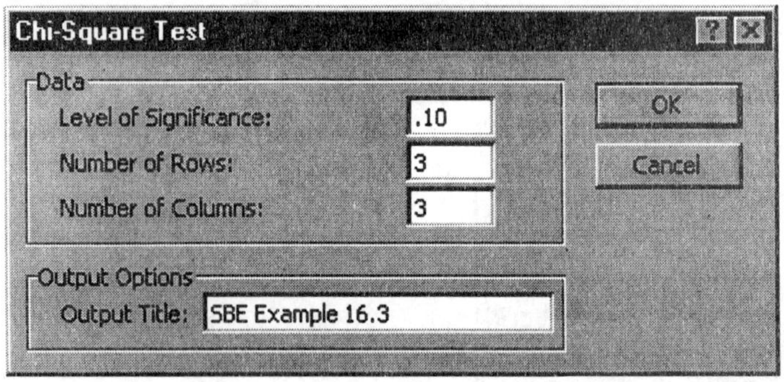

Section 16.2: Testing Categorical Probabilities: Two Way Table

Figure 16.3

SBE Example 16.3						
Observed Frequencies:			Column variable			
	Row variable	C1	C2	C3	Total	
	R1				0	
	R2				0	
	R3				0	
	Total	0	0	0	0	
Expected Frequencies:			Column variable			
	Row variable	C1	C2	C3	Total	
	R1	#DIV/0!	#DIV/0!	#DIV/0!	#DIV/0!	
	R2	#DIV/0!	#DIV/0!	#DIV/0!	#DIV/0!	
	R3	#DIV/0!	#DIV/0!	#DIV/0!	#DIV/0!	
	Total	#DIV/0!	#DIV/0!	#DIV/0!	#DIV/0!	
#DIV/0!						
Level of Significance	0.1					
Number of Rows	-0.5					
Number of Columns	0					
Degrees of Freedom	1.5					
Critical Value	2.7055406					
Chi-Square Test Statistic	#DIV/0!					
p-Value	#DIV/0!					
#DIV/0!						

This worksheet is a template that allows the user to change the values of the table cells to represent the data of their 3x3 table. We begin by replacing the generic labels **Column Variable** and **Row Variable** with the variable names of our example, **Client's Income** and **Broker Rating**. The next step is to replace the outcomes that are listed in the table as A1, A2, B1, and B2 with the outcomes that are meaningful in our example (e.g., Under $30,000, $30-$60,000, Over $60,000, Outstanding, Average, and Poor). The final step is to change the numbers shown in the table with the numbers that are shown in Table 16.1. The changed worksheet is shown in Figure 16.2

Figure 16.2

SBE Example 16.3

Observed Frequencies:

Broker Rating	Client's Income			Total
	Under $30,000	$30,000-$60,000	Over $60,000	
Outstanding	48	64	41	153
Average	98	120	50	268
Poor	30	33	16	79
Total	176	217	107	500

Expected Frequencies:

Broker Rating	Client's Income			Total
	Under $30,000	$30,000-$60,000	Over $60,000	
Outstanding	53.856	66.402	32.742	153
Average	94.336	116.312	57.352	268
Poor	27.808	34.286	16.906	79
Total	176	217	107	500

Level of Significance	0.1
Number of Rows	3
Number of Columns	3
Degrees of Freedom	4
Critical Value	7.779434
Chi-Square Test Statistic	4.2777051
p-Value	0.3697252
Do not reject the null hypothesis	

The chi-square test statistic and p-value found on this printout should be compared to the values found on the SAS printout on page 951 of the text. The test of independence can be conducted using this procedure for any two-way analysis regardless of the number of levels for each of the two classifications of interest.

Technology Lab

The following exercise from the *Statistics for Business and Economics* text is given for you to practice the categorical two-way procedure that is available within PHStat. Included with the exercise is the PHStat output that was generated to solve the problem.

16.38 An economist was interested in knowing whether sons have a tendency to choose the same occupation as their fathers. To investigate this question, 500 males were polled and each questioned concerning his occupation and the occupation of his father. A summary of the numbers of father-son pairs falling in each occupational category is shown in the table below. Do the data provide sufficient evidence at $\alpha = .05$ to indicate a dependence between a son's choice of occupation and his father's occupation?

Data

		Son			
		Professional	Skilled	Unskilled	Farmer
Father	Professional	55	38	7	0
	Skilled	79	71	25	0
	Unskilled	22	75	38	10
	Farmer	15	23	10	32

PHStat Output

SBE Exercise 16.38

Observed Frequencies:

Father	Son				
	Professional	Skilled	Unskilled	Farmer	Total
Professional	55	38	7	0	100
Skilled	79	71	25	0	175
Unskilled	22	75	38	10	145
Farmer	15	23	10	32	80
Total	171	207	80	42	500

Expected Frequencies:

Father	Son				
	Professional	Skilled	Unskilled	Farmer	Total
Professional	34.2	41.4	16	8.4	100
Skilled	59.85	72.45	28	14.7	175
Unskilled	49.59	60.03	23.2	12.18	145
Farmer	27.36	33.12	12.8	6.72	80
Total	171	207	80	42	500

Level of Significance	0.05
Number of Rows	4
Number of Columns	4
Degrees of Freedom	9
Critical Value	16.91896016
Chi-Square Test Statistic	180.874
p-Value	3.32841E-34
Reject the null hypothesis	